# Spotlight on Young Children and SCIENCE

Each issue of *Young Children,* NAEYC's award-winning journal, includes a cluster of articles on a topic of special interest and importance to the early childhood community. Most of the selections in this book originally appeared in *Young Children* Vol. 57, No. 5 in the cluster titled Teaching and Learning about Science. Others come from Beyond the Journal (www.naeyc.org), an online collection of resources that complement and expand on articles found in the journal, and from books that NAEYC has published or co-published.


Second printing 2004.

Cover photos (clockwise from top left):
Front cover/© Lois Main; © Kathy Sible; Herman Rich; courtesy Robin Friedrichs Moriarty; © Rick Toone, ETS; courtesy Patrica Howley-Pfeiffer. Back cover/ courtesy of Kathleen Conezio and Lucia French.
Illustrations throughout © Sylvie Wickstrom.

Through its publications program the National Association for the Education of Young Children (NAEYC) provides a forum for discussion of major issues and ideas in the early childhood field, with the hope of provoking thought and promoting professional growth. The views expressed or implied are not necessarily those of the Association. NAEYC thanks the contributors.

ISBN 1-928896-10-3
NAEYC #281

**Printed in the United States of America**

## Contents

# Spotlight on Young Children and SCIENCE

At first glance, teaching science in the early childhood classroom may seem daunting. Science is such a precise field that some teachers worry about their own capabilities in this area, let alone those of the children they teach. Yet there is no debate over whether science should be included in the preschool and primary curriculum. National and state standards stress that all children can learn science and that they have the right to become scientifically literate.

How can we resolve personal ambivalence with the child's right to learn science? If we move away from the traditional view of science as being all facts and memorization and instead look at it through our early childhood lens, we can see that for young children science is about trying to understand the world. Young children are natural scientists every day who observe the people, animals, and objects in their environment, conduct experiments, and record and report on their discoveries.

Science in early childhood is thus much broader than we might at first consider it to be. Let's avoid thinking about science in terms of activities devoted to "doing" science—such as growing mung beans or taking a nature walk. Rather, we should view science as an ongoing part of the total curriculum, woven into daily activities and routines. Science education occurs naturally when Samanda wonders why a cork floats and a penny sinks, when Juwan questions why popcorn kernels pop in a microwave, when Amy and Kirsten take apart a broken VCR, when John observes that he can hear his heart beating as he runs, and when Maria wonders why the water she paints on the blacktop disappears.

The view of the child as an active scientist has evolved since 1985 with the establishment of Project 2061, the American Association for the Advancement of Science's (AAAS) initiative for science reform in grades K–12. In 1989, Project 2061 published *Science for All Americans,* which defined the concept of science literacy. *Benchmarks for Science Literacy* outlined in 1993 what all children should be able to do in science by the end of grades 2, 5, 8, and 12. Drawing heavily on this content, in 1996 the National Research Council (NRC) defined standards for children at each grade level, from kindergarten through high school, in its document *National Science Education Standards.*

In 1998 the National Research Council and the American Association for the Advancement of Science jointly convened a technical forum on early childhood science, math, and technology and subsequently published *Dialogue on Early Childhood Science, Mathematics, and Technology Education*. All of these publications align with NAEYC's developmentally appropriate practice (Bredekamp & Copple 1997). They reinforce the idea that children can best learn science when it is presented through "hands-on," meaningful, and relevant activities.

The articles in this collection focus on innovative approaches educators are using to make early childhood science an exciting learning adventure. Many of the authors cite the AAAS and NRC documents in their approaches. The articles share the perspectives of teachers, staff developers, researchers, and scientists—all of whom are committed to teaching science to young children in this new, exciting way.

"Science in the Preschool Classroom: Capitalizing on Children's Fascination with the Everyday World to Foster Language and Literacy Development," by **Kathleen Conezio** and **Lucia French,** underscores the point that science for young children ought to focus on the world in which children live. The authors view science as part of an integrated curriculum rather than isolated subject matter and make a case for using science as a foundation for teaching language and literacy skills. Included with this article are samples of a planning wheel and a science unit on liquids that integrates in-depth exploration of a science phenomenon with literature and language.

"Entries from a Staff Developer's Journal: Helping Teachers Develop as Facilitators of Three- to Five-Year-Olds' Science Inquiry" is by **Robin Moriarty,** a staff development specialist. She kept a journal of her observations as teachers conducted their own science explorations.

"Using Photographs to Support Children's Science Inquiry," by **Cynthia Hoisington,** describes the use of digital photography to help children become astute observers and problem solvers.

In "Documenting Early Science Learning," **Jacqueline Jones** and **Rosalea Courtney** encourage teachers to collect, describe, and interpret evidence of young children's emerging science understandings so they can better plan science instruction. The authors urge teachers to collect evidence over time for individuals and the group, and offer examples of various kinds of documentation.

"Be a Bee and Other Approaches to Introducing Young Children to Entomology," by **James Danoff-Burg,** describes how to involve children in the study of insects by making live collections of insects, so children can conduct real-life investigations. He also describes how children can construct insect models and take on the insect's perspective through a movement activity.

Sometimes teachers have a special interest to share with young children. In "Raising Butterflies from Your Own Garden," **Patricia Howley-Pfeifer** describes how her passion for butterflies enhanced the curriculum and encouraged children to observe, keep detailed records, and communicate their scientific discoveries through drawing, writing, and dictations about what they learned.

*—Laura J. Colker and Derry Koralek*

# Science in the Preschool Classroom

# Capitalizing on Children's Fascination with the Everyday World to Foster Language and Literacy Development

Kathleen Conezio and Lucia French

A young child starting preschool brings a sense of wonder and curiosity about the world. Whether watching snails in an aquarium, blowing bubbles, using a flashlight to make shadows, or experimenting with objects to see what sinks or floats, the child is engaged in finding out how the world works.

It is not exaggerating to say that children are biologically prepared to learn about the world around them, just as they are biologically prepared to learn to walk and talk and interact with other people. Because they are ready to learn about the everyday world, young children are highly engaged when they have the opportunity to explore. They create strong and enduring mental representations of what they have experienced in investigating the everyday world. They readily acquire vocabulary to describe and share these mental representations and the concepts that evolve from them. Children then rely on the mental representations as the basis for further learning and for higher order intellectual skills such as problem solving, hypothesis testing, and generalizing across situations.

While a child's focus is on finding out how things in her environment work, her family and teachers may have a somewhat different goal. Research journals, education magazines, and the popular press are filled with reports about the importance of young children's development of language and literacy skills. Children's natural interests in science can be the foundation for developing these skills.

**Kathleen Conezio,** M.S., is director of curriculum and professional development for two education grants through the University of Rochester. Kathleen has more than 20 years of experience as a teacher and education coordinator in private and public preschools and in Head Start. She is co-author of the ScienceStart! Curriculum.

**Lucia French,** Ph.D., is a developmental psychologist on the faculty of the Warner Graduate School of Education and Human Development at the University of Rochester. Lucia's areas of basic research include young children's language and cognitive development. She is responsible for the creation and field testing of the ScienceStart! Curriculum.

Funding for the ScienceStart! Curriculum comes from the National Science Foundation (Award ESI-9911630), U.S. Department of Education (Award S349A010171), the A.L. Mailman Family Foundation, and Rochester's Child.

Photos courtesy of the authors.

Back in February, Mrs. O'Shea's preschool children had explored the concept of light and shadows. They collected many types of materials to see which ones would create a shadow in the bright light and which ones the light would just pass through. After several days of experimentation, they realized that while opaque materials create shadows and transparent materials allow light to pass through easily, there are some things that don't fit either category. These materials allow some light to pass through (although not as much as window glass) and they cause very light shadows. Later in the school year, a visitor to the classroom was present during snack time when the children were trying new clear strawberry flavored Jello with stars and moon shapes in it. The visitor overheard the following conversation among the four-year-olds:

"It's transparent!" remarked one little girl with surprise.

"No, it's translucent," countered another girl.

"Why do you say it's translucent?" asked Mrs. O'Shea.

"Because you can only see through it a little," the girl responded.

Whereas many adults think of science as a discrete body of knowledge, for young children science is finding out about the everyday world that surrounds them. This is exactly what they are interested in doing, all day, every day.

In the preschool classroom or in the university research laboratory, science is an active and open-ended search for new knowledge. It involves people working together in building theories, testing those theories, and then evaluating what worked, what didn't, and why.

On a bright fall morning, a group of three-year-olds takes a walk and observes fall leaves dropping from the trees and blowing around the school yard. They come inside to read a book. The book contains a picture of a rake. Few of these urban, apartment-dwelling children have ever actually seen a rake. The teacher asks what it is and what it might be used for. A real rake is brought in as the discussion proceeds and the children speculate:

"You could scratch the grass."

"Use it for a back scratcher."

"Throw it in the garbage."

"I clean the leaves!"

All of the children's ideas are considered and a bag of fall leaves is dumped on the classroom floor. The children are given opportunities to feel the leaves, kick the leaves, and use the rake. Coming back together as a group, they reevaluate their earlier theories and decide that a rake can be very helpful in making a pile of leaves to jump into. Science, language, and literacy have all combined in a meaningful learning experience for the children.

## A science-based curriculum

For the past seven years, the authors have been involved in creating, implementing, and refining a science-based preschool curriculum encompassing both content and process goals. Known as ScienceStart! this full-day, full-year curriculum is currently being field-tested at a number of sites in Rochester, New York. (Throughout, when we refer to one of "our" classrooms, we mean a classroom that is using ScienceStart!) The major content goal of this curriculum is for children to develop a rich, interconnected knowledge base about the world around them. The primary process goal is to foster and support the types of typical intellectual development that characterize the preschool years. These include receptive and expressive language skills, skills in self-regulation—particularly attention regulation—and skills in problem identification, analysis, and solution. Several theoretical assumptions that are widely shared by early childhood professionals underlie these goals:

- Young children are active, self-motivated learners who learn best from personal experience rather than from decontextualized linguistic input (e.g., French 1996; Nelson 1996).
- Young children construct knowledge through participation with others in activities that foster experimentation, problem solving, and social interaction (Gallas 1995; Chaille & Britain 1997).
- Young children should be allowed to exercise choice in the learning environment (Bredekamp & Copple 1997).
- Children's social skills develop best when they have opportunities to learn and practice them in the context of meaningful activities (e.g., Katz & McClellan 1997).

Science in our preschool classrooms is not a complicated process, nor is it an activity that occurs separately from the normal classroom routine. Almost all young children in almost all environments "do science" most of the time; they experience the world around them and develop theories about how that world works.

Almost all young children in almost all environments "do science" most of the time; they experience the world around them and develop theories about how that world works.

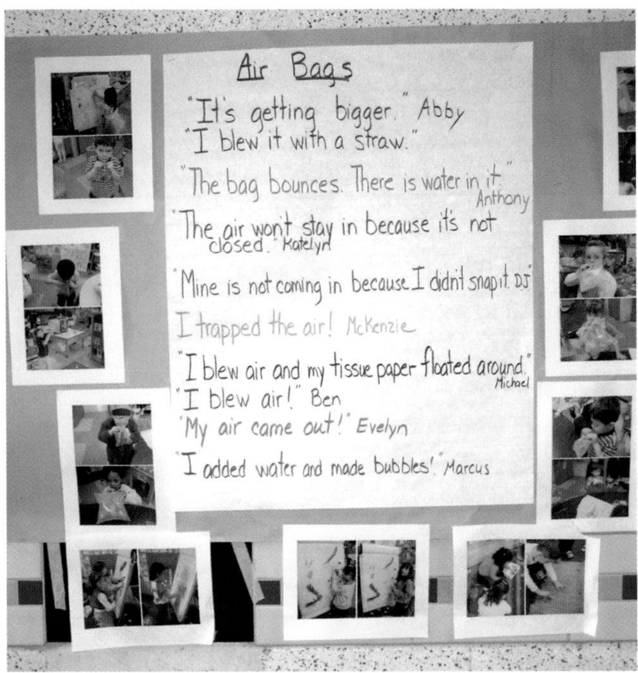


At the easel, a boy is using blue and yellow paint. Suddenly, he notices that as he paints, the color green appears. The child has the opportunity to theorize about color mixing: "Does this always happen with blue and yellow paint?" "Can I make any other colors with blue and yellow paint?"

In any preschool classroom, the process of formulating theories based on experience happens in the art, block, and dramatic play areas, and during outdoor play. The difference for the children in our classrooms is that adults work to create an environment that is integrated and coherent rather than disjointed. Thus, children explore the same phenomenon—in this case, color mixing—in different parts of the classroom, particularly in activities that involve language and literacy.

**Real science begins with childhood curiosity, which leads to discovery and exploration with teachers' help and encouragement.**

In the leaf raking example described earlier, the children took a walk outside to see leaves blowing, then read a story about leaves, then raked leaves in the classroom. They also had other opportunities that day to explore leaves. They could decorate leaf-shaped cookies with a variety of fall-colored frosting, paint tree and leaf pictures at the easel, sort real leaves by shape or color, examine leaves with a magnifying glass, and dance like leaves in the wind. Science, for all the children, was a creative and exploratory process, one in which they could use many forms of knowledge to build theories about their world. While talking with them about what they were doing, the teacher not only involved the children in a conversation, but also offered them relevant vocabulary and modeled ways of thinking about and talking about their experiences.

## Childhood curiosity leads to real science

Many early childhood teachers are hesitant about introducing science in their classrooms, often because of their own unpleasant science education experiences. When asked if they teach science, these educators might point to the plants on the shelf or the collection of stones and shells and indicate that science is taking place "over there." Other teachers see science as some kind of magic trick to perform on a Friday afternoon when everyone is tired and bored. They bring out the baking soda and vinegar to "make a volcano." While the children may be amazed and amused by this activity, it does not build accurate knowledge and does not represent real science.

Real science begins with childhood curiosity, which leads to discovery and exploration with teachers' help and encouragement. It involves three major components: content, processes, and attitude. Young children prize information about the world around them, yet an emphasis on content is not enough. Although many people view science as a body of knowledge (facts and formulas) that scientists learn and use, in reality this body of knowledge is constantly changing as new discoveries are made. Young children, like scientists, need to practice the process skills of predicting, observing, classifying, hypothesizing, experimenting, and communicating. Like adult scientists, they need opportunities to reflect on their findings, how they reached them, and how the findings compare to their previous ideas and the ideas of others. In this way, children are encouraged to develop the attitude of a scientist—that is, curiosity and the desire to challenge theories and share new ideas. Scientific exploration presents authentic opportunities to develop and use both receptive and expressive language skills.

In Miss Chrissie's classroom, one morning in April, an observer asked two four-year-old girls what was inside the cups on a windowsill. The girls explained that they had planted seeds and were waiting for them to grow. The observer asked how long it would take and was told, "Maybe a few days." The observer asked why it would take so long and was told, "Growing takes time. You need to be patient." The girls then explained about the plant's need for water and light. The observer looked outside and pointed out to the girls that there were grass, trees, and flowers outside that also needed water. The girls reassured her that the rain would water those plants. While this may appear to be a simple and everyday conversation (as indeed it should be), these girls were using their classroom science work to make observations and hypotheses and communicate these clearly to a classroom visitor.

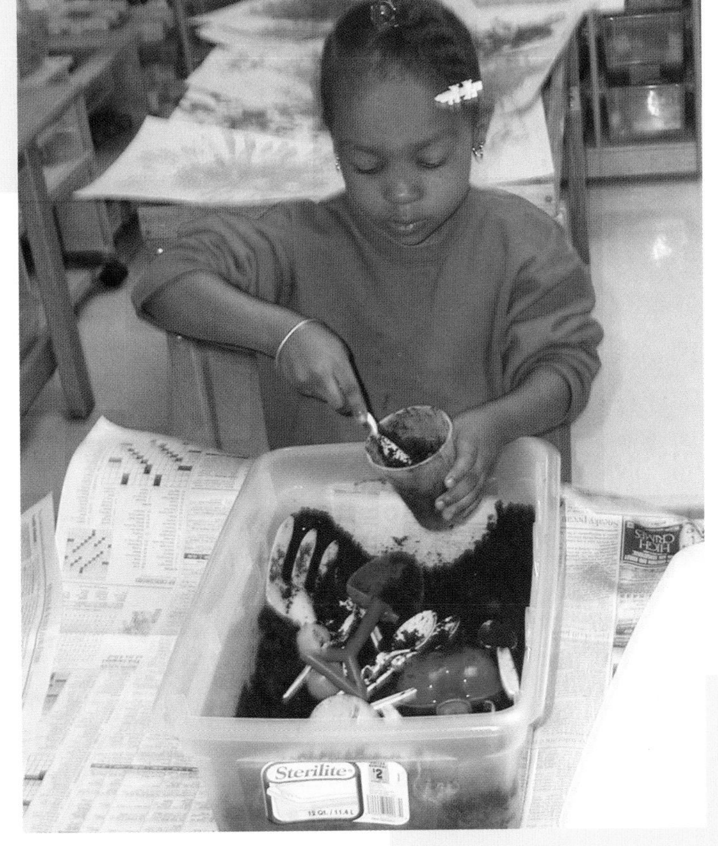


## The importance of a coherent approach

In *Talking Their Way into Science,* Karen Gallas (1995) explains that young children must be allowed to co-construct their knowledge about science by imagining possible worlds and then inventing, criticizing, and modifying those worlds as they participate in hands-on exploration. They must be encouraged to develop possible theories about their own questions and then proceed to investigate these theories within the classroom learning community. For this to happen, the opportunity for in-depth and long-term investigation through a variety of activities—what we term *coherence*—is essential. The girls whose seeds were growing on the windowsill had opportunities to over- and under-water plants; paint bouquets of flowers at the easel; take plants apart to investigate the roots, stems, and leaves; and make and eat a salad containing leaves, roots, stems, and flowers. They read many books about plants and participated in discussions with peers and adults about what they were learning.

Many, and perhaps most, preschool classrooms have little coherence from day to day. For example, teachers following a "letter of the week" approach may have children investigate dinosaurs one day, dig in dirt the next day, and make a dessert the third day. Each activity is developmentally appropriate and enjoyable, but other than the letter *D* they have nothing in common.

In contrast, in a coherent approach to early childhood education, each day's activities build on those of the day before and provide a basis for those of the following day. Teachers who follow a science-based curriculum find that they can maintain a focus for 8, 10, or even 12 weeks. For example, the ScienceStart! unit on color and light takes place over a 10-week period. Children explore mixing colors to make new colors, investigate light sources and how shadows are made, observe how light travels, and finally study the cycle of day and night. While each day brings new activities and new theories, the days fit together into a coherent pattern that offers children the opportunity to revisit ideas and activities, to build a knowledge base, and to use knowledge gained on one day as the foundation for the next day's exploration.

It might seem that learning about air could be difficult for four-year-olds. After all, they can't see it or even really get ahold of it. But we have found that after spending the previous eight weeks discovering the properties of solids and liquids, preschool children have a lot to say about air.

"I know it's there 'cause I can feel it in my hair."

"The bubble has my air in it!"

"Air isn't like a solid 'cause it has no shape. It's the shape of the balloon."

"You can't pour it and it doesn't make a mess on the floor."

## Science at the Center of the Integrated Curriculum: Ten Benefits Noted

**1. Science responds to children's need to learn about the world around them.** The primary reason for a science-based early childhood curriculum is that children love it. Disruptive behavior diminishes as children become engaged in explorations. Conversation and cooperation increase as children talk with one another about their predictions, observations, and questions.

**2. Children's everyday experience is the foundation for science.** For example, the concept and process of "change" can be explored by making scrambled eggs. Children know what raw and scrambled eggs look like, but may have never watched the transformation take place. Focusing on the process of cooking the egg offers a new way of considering familiar objects and events and provides a meaningful context within which to introduce new vocabulary and science concepts.

**3. Open-ended science activities involve children at a wide range of developmental levels.** Within any classroom activity, there are a variety of levels at which children can engage, depending on their prior knowledge and skills. For example, when using water-droppers to mix colored water, one child may spend 20 minutes practicing the small motor skills for operating a water-dropper while another child spends an equal amount of time exploring how to use proportions to create different shades of orange. Because children can find their own level within an activity, they are challenged without becoming frustrated or bored.

**4. Hands-on science activities let teachers observe and respond to children's individual strengths and needs.** As teachers observe children finding their own level within an open-ended activity, they can become more aware of what the child knows and what she may need some assistance with. For example, the child who practiced small motor skills with a water dropper may enjoy more tasks to strengthen that skill or may be ready to repeat the activity at a higher conceptual level (that is, focusing on creating colors).

**5. The scientific approach of "trial and error" welcomes error—interprets it as valuable information, not as failure.** Achievement increases when children are free to focus on learning rather than on avoiding mistakes.

**6. Science strongly supports language and literacy.** Children learn language through participation in meaningful, comprehensible language-based interactions. Appropriately implemented, a science-based curriculum is rich in language use by both adults and children. Literature of all kinds can be used to support a science-based curriculum. Songs, finger-plays, poems and books can be matched to the activity and used to support it.

A. Nonfiction books become a powerful foundation for conversations with adults and peers (Look what the inside of a frog looks like! How can people stay warm if they live in an igloo?).

B. Vocabulary growth is supported by children's prior knowledge/experience of the everyday world, coupled with observation and hands-on activities. For example, a child who has watched her father make scrambled eggs has prior knowledge that she can draw on to interpret a hands-on classroom experience of cracking and heating eggs. This knowledge and the present experience provide a meaningful situation to support the child's learning of new vocabulary, perhaps words such as *raw, yolk, stir, cooked, spatula, heat,* and *change.*

C. Receptive language (listening comprehension) is fostered as children listen to the teacher read aloud and talk about the science activity.

D. Expressive language is fostered as the teacher leads children through a cycle of scientific reasoning and especially as the teacher supports the children in developing a report of their findings.

**7. Science helps English-language learners to participate in the classroom and learn English.** Teacher demonstrations and hands-on activities with familiar materials enable children who come from a home where English is not spoken to understand a great deal of the content without understanding the teacher's language. Their understanding of the situation helps them learn English.

## y Head Start Teachers

**8. The problem-solving skills of science easily generalize to social situations.** Teachers can help children adapt the cycle of problem solving to interpersonal problems. They can help children to plan some possible solutions and to predict what might work best, then encourage them to try the proposed solution and let the teacher know how it worked, and then help them try something else if the first attempt didn't work.

**9. Science demonstrations help children become comfortable in large-group conversations.** When the teacher makes orange by combining red and yellow, children are amazed and ask how and why it happened and what would happen if other colors were mixed. The teacher can support and extend a large group conversation of this sort for several minutes, and then suggest ways for children to explore the questions they have generated. When demonstrations and discussions take place in a large group setting, the children all share the same experience and knowledge base; this creates a community of learners who can support one another's explorations, share new ideas on a topic, and challenge new theories generated.

**10. Science connects easily to other areas, including center-based play, math, artistic expression, and social studies.** With an integrated curriculum, related activities/concepts are explored in several locations in the classroom. This offers children an opportunity to learn using a variety of different senses and skills. For example, during a unit on color mixing, children might find net capes in primary colors in the housekeeping corner (these can be layered to create secondary colors), two primary colors at the easel with a model of various shades of the resulting secondary color, and, at the science table, strips of colored cellophane and clear contact paper to create plaid sun-catchers.

While children's theories are seldom complete and will go through many revisions, the coherence of the curriculum offers them opportunities to make in-depth explorations over an extended period of time.

### Science learning: Something to talk about

Several years ago, the local director of state-funded preschool programs was asked why she was spending money on inservice training in the area of science when, after all, "everyone knows" that language and literacy should be the focus during preschool. Agreeing that language and literacy were important goals for young children, the administrator pointed out that language and literacy learning must be *about something.*

After hearing this story, we asked our teachers, who had been using a science-based curriculum for several years, to respond to the Why science? question based on their own observations and experiences. The resulting conversation was condensed into the 10 good reasons (shown at the left in "Science at the Center of the Integrated Curriculum: Ten Benefits Noted by Head Start Teachers").

There can be many reasons for a science focus in the preschool years. Because science is so intriguing for young children, they become more engaged and therefore more attentive to and involved in the language of the classroom. A coherent, integrated curriculum allows for more complex language use and more sustained literature studies than does a disjointed approach to content.

Teachers may wonder how language and literacy experiences are integrated into a science-focused curriculum. Researchers have found that children are most likely to learn language and literacy skills when they have opportunities to use these skills in authentic situations (e.g., Goodman 1984; Teale & Sulzby 1984). The problem-solving approach associated with scientific inquiry is rich in language. Teachers can support children as they acquire and practice increasingly sophisticated language skills. The group discussion may be completed in 5 minutes or may continue as long as 45 minutes. Throughout this period, participants are involved in coherent, contingent conversation. Whether active contributors to the conversation or listeners, children gain important practice in how to maintain conversational coherence, switch and return to topics, use language to move among the past, present, and future, and translate between linguistic and mental representations.

To speak, children must translate their own mental representations into linguistic output that can be shared with others. In listening, they create mental representations based on someone

Because science is so intriguing for young children, they become more engaged and therefore more attentive to and involved in the language of the classroom.

else's language. Translation between linguistic form and mental representation is generally difficult for young children, but in this case it is supported and facilitated by the hands-on experience being shared by the listener and speaker.

> As children were gathered around the duck egg incubator in Mrs. Toot's classroom, the teacher asked them what they knew about ducks. The children speculated about what ducks eat, and asserted that ducks quack and can swim. One girl added that they have "skin between their toes." The discussion continued about what covered their bodies, with some children arguing that it was fur, while others contended that feathers cover a duck. No agreement was reached, and the suggestion was made that they needed a real bird to look at. Mrs. Toot arranged a classroom visit from a parakeet while they waited for their duck eggs to hatch.

Investigations of the everyday world offer many opportunities for a variety of preliteracy and literacy experiences. There are opportunities for receptive and expressive language, for consulting text, and for producing graphic representations of ideas (both drawn and written). So, in our classrooms, the daily literacy activities are integrated into the science learning. As in many other preschool classrooms, our science-focused teachers read to their children every day.

Children work together to create written reports about their scientific explorations. They make graphs and charts, create books, and dramatize ideas. Many children keep science journals to record data. For example, in our classrooms three- and four-year-old children from families living in poverty used drawings and words to document the growth and changes that occurred as their caterpillars transformed into painted lady butterflies. Strong and meaningful learning takes place as children participate in language and literacy experiences about something of real significance to them.

**Strong and meaningful learning takes place as children participate in language and literacy experiences about something of real significance to them.**

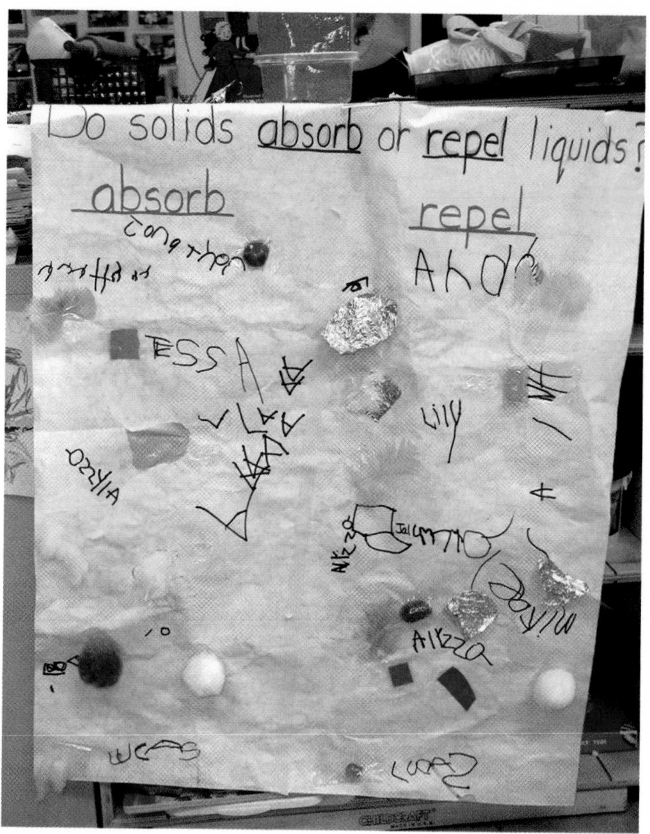


## Conclusion

Some teachers want to take steps to introduce more science into their education programs, but they are unsure about what to do. These same teachers are often comfortable with cooking and art. It is possible to explore many science activities through cooking and art. A coherent unit can be developed in which the same topic is explored through three activities—science, art, and cooking. For example, the effects of air could be explored by making meringue cookies (cooking), by using a straw and hair dryer to blow a marble across a page containing wet paint to create an air picture (art), and by taking a collection of items and predicting which can be moved by blowing through a straw (science).

Teachers who increase their understanding of what science is at the preschool level will come to see that science can be incorporated into many, if not most, of the activities that they already do. Science itself is not an activity, but an approach to doing an activity. This

## Using Language during Science Activities

In science activities with young children, we follow a simplified four-step cycle of scientific reasoning. This format is repeated daily as each new activity is introduced. As children internalize the format, it becomes an automatic and powerful guide for problem solving in many situations, not just science activities. The forms of language use associated with each of the four components of the cycle of scientific reasoning are described below.

### Ask and reflect

The cycle of inquiry begins with questions: "I wonder what would happen if . . . ?" "I wonder why . . . ?"

Most young children are curious about the world around them, but many have not had much experience asking questions. They may need adult modeling and support for a while, but they will quickly learn how to ask open-ended questions.

Whether a teacher or child has provided a question, once it is on the table it is time to reflect on what is already known that might relate to the question. This is an opportunity for children to think and then to translate their thoughts into language.

It might also be appropriate to read aloud one or more books to learn more about the topic. Listening to the teacher read aloud provides children with the opportunity to create mental representations or knowledge from linguistic input. Children enjoy being read to under almost any circumstance, but listening comprehension is greatly facilitated when they have specific questions.

### Plan and predict

When children have a question and have considered what related information they already know, it is time to plan how to address the question. When we see people act, we usually don't know whether or not they have a plan and are following it. This is because planning usually takes place silently.

To learn how to plan, children need to see other people planning. It is important that teachers show children their planning process. The teacher can elicit a plan from children with careful questioning or can propose plans herself for the children to evaluate. Again, this process involves translating back and forth between linguistic representations and mental representations.

After a plan has been formulated, children need to make predictions about what the outcome will be. Different children will have different predictions. The teacher should elicit these predictions in a way that helps children think carefully but that does not make them worry about whether or not they are correct.

One of the most important functions of language is *displaced reference,* the ability to use language to refer to things that are in the future or past or that are in other locations. Planning and predicting both involve displaced reference. Practice in this type of language increases children's discourse competence.

### Act and observe

Finally, it is time to put a plan into action. This is the hands-on part that children enjoy so much. After a plan has been carried out, children can observe the results of the action and compare what actually happened to their predictions. Again, there are opportunities for expressive language (describing what happened, describing the match between prediction and finding) and for receptive language (listening to other participants' descriptions). Children learn to make and evaluate explanations.

### Report and reflect

Sharing observations with others is an important part of the scientific cycle. There are many different types of reports. To share and display their findings, children can tell someone, dictate text (individually or as a group), draw or make a chart or graph, or even put on a skit or create a song. All these forms of reporting involve authentic language and literacy opportunities. The reflections at the end of the cycle are likely to set the stage for reflections at the beginning of the next day's activity.

approach involves a process of inquiry—theorizing, hands-on investigation, and discussion.

Over the past seven years, we have worked with almost two dozen teachers who were implementing ScienceStart! predominantly with children from families with low incomes, including children with special needs. We have found consistent reactions among these teachers. They find that an emphasis on hands-on science leads to increases in children's level of engagement, in language use and language skills, and in positive peer interactions. Families have been surprised by their children's abilities to learn science, and report that their children often transfer content knowledge and the process of inquiry from preschool to the home environment. For example, while in the backyard with his mother, one three-year-old asked, "What do you think

*(Continued on p. 14)*

## Quick Recipe Science Unit

Some early childhood teachers are enthusiastic about jumping right in and implementing an all-day, every-day science-based curriculum. Others are more cautious, yet appreciate the importance of providing children with a coherent approach that will support the development of a rich knowledge base. For these teachers, we have developed what we call the **Quick Recipe**.

Each Quick Recipe integrates literature, language use, and extended exploration of a particular phenomenon. The lesson may take only 45 to 90 minutes or may extend over a three- to five-day period. **Ingredients include a topic, two books related to the topic, three activities, and several open-ended questions about the topic.**

For example, if the children are interested in insects, the teacher might introduce the topic by reading a book such as *Bumblebee, Bumblebee, Do You Know Me?* by Anne Rockwell or might talk with the children about the topic prior to reading aloud. Together, the teacher and children can generate some questions, or "wonder," about the topic. The children may ask what kinds of insects there are, what insects like to eat, or can they all fly.

Using the children's questions and her own ideas as a basis, the teacher can plan a set of interrelated activities. She can introduce each activity in a large or small group setting; then, in a small group setting, she can support children as they work through the science cycle (plan and predict, act and observe, report and reflect). Note: If the teacher offers all activities on the same day, with children rotating from one activity to the next, it is important to have an adult available to support each activity.

For an insect exploration, activities might include an outdoor hunt for insects, observing the insects with magnifying glasses and drawing them, and feeding insects in a terrarium. After all the children have participated in the three activities, the large group can reconvene and discuss what was learned. This learning can be related to the questions generated and written down during the earlier discussion.

Then the teacher can read and discuss another book, such as *Are You a Ladybug?* by Judy Allen and Tudor Humphries. At this point, she may feel there is enough interest and curiosity for the class to continue exploring the topic, or she may decide to move on to a different one.

A sample Quick Recipe for Liquids follows. Using this same format, teachers can readily create their own Quick Recipes in response to children's interests.

## Quick Recipe for Liquids

**Two books**

*Wet or Dry* by Bruce MacMillan
*Down Comes the Rain* by James Hale

**Two guiding questions**

What are the characteristics of liquids?
What liquids do we have around us?

**Three activities**

Moving Water
Varieties of Liquids
Magic Touch Bags

### Moving Water

#### Learning Objective

Water can move and be moved in many different ways. Water does not have a shape of its own. It takes the shape of whatever you put it into. If you just drop water onto a surface, it makes a puddle.

#### Materials Needed

baster • straws • eyedroppers • wide-mouth containers • dishpans • wax paper • towels for spills • cookie sheets (for a ramp) • water

#### Experiment

**1.** Set up three water stations with different tools: (a) baster, (b) straws, and (3) eyedroppers. Make a ramp by propping up against a stack of blocks one end of a cookie sheet covered with wax paper. Have the end of the cookie sheet drain into a dishpan.

**2.** Take water out of a container with one of the tools and release it down the ramp into the dishpan. Change the height of the ramp by adding or removing blocks.

**3.** Observe how the water moves.

#### Discovery Questions

**1.** What tool moved the water the fastest?

**2.** How long did it take to move the water from the container using an eyedropper? the baster? a straw?

**3.** Describe what the water looked like when it went down the ramp. Did it move quickly or slowly?

**4.** What other tools could we use to move water?

**5.** What would have happened if you had tried to do this with wood blocks instead of water? Would they have moved the same way?

#### Connections

**1.** Make different size drops of water on the waxpaper lying flat.

**2.** Measure the height of the ramp needed to make the water flow faster.

3. Make a plan to watch the weather and the effects of rain outside.
4. Sing *Rain, Rain, Go Away.*
5. Make a rainstick musical instrument.
6. Play with a marble maze.

### Follow-up Book

*Rain* by Robert Kalan or *The Rain-Player* by David Wisniewski

## Varieties of Liquids

### Learning Objective

There are many different kinds of liquids in our world. In some ways they are all the same—they all take the shape of their container, for example. In other ways, they can be different. We can find out about these similarities and differences by doing things with liquids and making observations.

### Materials Needed

baster • straws • water • milk • juice • molasses • cooking oil • eyedroppers • wide-mouth containers • paint • dishpans • wax paper • dish soap • towels for spills • cookie sheets (for a ramp)

### Experiment

1. Set up three water stations as in the previous activity.
2. This time have a variety of liquids at each station.
3. Move liquids with the various tools and observe what happens.
4. Think about how you will remember and keep track of what happens.

### Discovery Questions

1. Are all liquids wet?
2. Do all liquids flow down the ramp? Do they all flow the same way? Describe what happened.
3. Why do some liquids flow quickly and others flow very slowly?
4. Which liquids did you try today that were most like water?
5. What other liquids could you try?
6. What would happen if you mixed them all together?

### Connections

1. Measure the length of a drip on the ramp made with various liquids. Compare.
2. Make droplets out of various liquids. Describe the droplets by using *small, medium,* and *large.*
3. Discuss the uses of each of the liquids. Does the use have anything to do with the characteristics?
4. Act out a thunderstorm.
5. Do straw painting.
6. Go outside and walk and jump through puddles.
7. Make Kool-Aid and lemonade.

### Follow-up Book

*The Milkmakers* by Gail Gibbons or *Lulu's Lemonade* by Barbara Derubertis

## Magic Touch Bags

### Learning Objective

Liquids may mix together very well or may separate when they are mixed.

### Materials Needed

heavy-duty sandwich-size zipper bags (one per child) • cornstarch • water • oil • food coloring or dry tempera paint

### Experiment

1. Combine 5 Tbs. of cornstarch, ½ cup of water, and 5 or 6 drops of food coloring in each zipper bag. Gently mix.
2. Add ½ cup of oil to the bag.
3. Seal the bag tightly.
4. Lay the sealed bag flat on the table and press the ingredients in the bag with a finger.
5. Watch for the colors to mix and separate.
6. Shake the bag gently in front of you and watch what happens.

### Discovery Questions

1. How did you make the liquids move?
2. What happened to the color in the liquids when you pressed with your finger or shook the bag?
3. Why didn't the oil and water stay mixed together?
4. Where did the oil go when they separated? Did this happen every time you mixed them?
5. What do you think would happen if you used milk instead of water? Milk instead of oil? Try this and see what happens.

### Connections

1. Count and measure all ingredients.
2. Identify colors when the liquids mix together.
3. Play Laundromat in the housekeeping area.
4. Wash clothes in the water table using soap and water.
5. Sponge paint.

### Follow-up Book

*Mouse Paint* by Ellen Stoll Walsh

will happen if we add water to this dirt? What do you think we will get?"

In 1993 the American Association for the Advancement of Science published *Benchmarks for Science Literacy,* a compendium of specific science goals for K–12 grade levels. The use of a coherent, hands-on science curriculum provides preschoolers with the opportunity to meet virtually all of the benchmarks described for children in the K–2 range. For example, at a very general level, the benchmarks for kindergarten through second grade are as follows:

> Students should be actively involved in exploring phenomena that interest them both in and out of class. These investigations should be fun and exciting, opening the door to even more things to explore. An important part of students' exploration is telling others what they see, what they think, and what it makes them wonder about. Children should have lots of time to talk about what they observe and to compare their observations with those of others. A premium should be placed on careful expression, a necessity in science, but students at this level should not be expected to come up with scientifically accurate explanations for their observations. (AAAS 1993, 10)

Most young children bring curiosity and wonder to the early childhood setting. Teachers need only capitalize on these characteristics to make science learning come alive every day. Science learning provides a rich knowledge base that will become an essential foundation for later reading comprehension. It also provides the foundation for meaningful language and literacy development.

## Familiar Children's Books That Can Be Related to Science Topics

**Air:** *How Does the Wind Walk?* by Nancy Carlstrom; *Air Is All Around You* by Franklyn Branley; *Gilberto and the Wind* by Marie Hall Ets

**Liquids:** *Lemonade for Sale* by Stuart Murphy; *Rain* by Robert Kalan; *Bringing the Rain to Kapiti Plain* by Verna Aardema

**Color:** *Mouse Paint* by Ellen Stoll Walsh; *Skyfire* by Frank Asch; *The Color Box* by Dayle Ann Dodds

**Shadows:** *Shadow Night* by Kay Chorao; *Bear Shadow* by Frank Asch; *What Makes a Shadow?* by Clyde Robert Bullo

**Living things:** *Feathers for Lunch* by Lois Ehlert; *A Field of Sunflowers* by Neil Johnson; *How Have I Grown?* by Mary Reid

To receive an extensive list of fiction and nonfiction children's books for use in science units on properties of matter (liquid, solid, gas, and change) and color and light, please write: **Dr. Lucia French, Warner School, University of Rochester, Rochester, NY 14627.**

## Color and Light Integrated

**A Color and Light Integrated Planning Wheel** is an example of the planning format we developed to help teachers create an integrated curriculum. The lead activity of the day, "Colorful Clay," is part of a coherent unit on mixing colors.

The teacher introduces the topic in a large group setting and reads the book *Little Blue, Little Yellow* by Leo Lionni to engage children in a discussion of the topic. During large group time, the teacher might demonstrate—while inviting lots of input from the children—the entire activity. Next, during choice time, children can extend or repeat the science activity in a small group with adult support. Children who choose to play in other activity settings will find props and materials to support further exploration of the activity.

The listing of related math and social studies activities and relevant vocabulary in the planning wheel helps the teacher extend the day's lesson.

### Colorful Clay Unit

#### Concept

We can identify the primary colors: red, yellow, and blue. In combinations, they create the secondary colors: green, orange, and purple.

#### Objectives

The children will

- ask and reflect about the colors they see around them every day
- plan and predict how to make more colors
- act and observe what happens when the primary colors are mixed together
- report and reflect on the new colors they have created.

#### Materials Needed

red, yellow, and blue playdough • plastic wrap • plastic knives • red, yellow, and blue crayons • paper

## References

American Association for the Advancement of Science (AAAS). 1993. *Benchmarks for science literacy: Project 2061.* New York: Oxford University Press.

Bredekamp, S., & C. Copple, eds. 1997. *Developmentally appropriate practice in early childhood programs.* Rev. ed. Washington, DC: NAEYC.

# Planning Wheel

Color and Light
Integrated Planning Wheel

Colorful Clay

Unit 1 - Following the questions

Science Activity — Colorful Clay

Unit 1 Color and Light — Following the Questions

**Social Studies**
Skin color - the colors of friends in our world.

**Language Arts**
Vocabulary: Playdough, Mix, Knead, New, Compare, Separate
Literature: *Little Blue, Little Yellow* Leo Lionni
Expressive Language: Retell the story, *Little Blue, Little Yellow* in your own words
Receptive Language: With teacher's help, follow recipe to make playdough

**Mathematics**
Collect and count blue, yellow, and green objects in the classroom.

**Arts & Expression**
Easel painting
chromatography garden
"Little Boy Blue" Horns
Stringing fruit loop necklaces
Song - "Little Red Box"
Playdough
Write a class book – Brown Bear, Brown Bear...
Sponge painting
Marble painting

**Outdoor Play**
Hide and seek for colors
Traffic lights in the neighborhood
Play "red light - green light"
Colors on a neighborhood walk

**Center Based Play**

Water Table: Eyedroppers, Colored water, Making colorful floats

Block Area: Duplo Blocks, Street signs, Windows for block buildings

Manipulatives: Snap blocks, Montessori Sorting box, Jelly bean sorting, Pegs & pegboards

Science Table: Color bottles, Color paddles, Paint store pallette, Prisms

Housekeeping Area: Dress up clothes, Fruits and vegetables, Placemats & tablecloths

Reading Corner: ***Color Dance,*** Ann Jonas; ***The Color Box*** Dayle Ann Dodds; ***Freight Train,*** Donald Crews; ***If You Take a Print Brush*** Fulvio Testa; ***The Mixed Up Chameleon*** Eric Carle

## Activity

*Be sure to give the children plenty of time to experiment with the playdough before doing this experiment.*

1. Choose one of the colors of playdough and roll it into a long rope.
2. Select a crayon that matches the color of the dough.
3. Draw on paper a mark to symbolize the dough.
4. Repeat with a second color.
5. Slice off equal amounts of the two colors that were chosen and predict what new color will be created.
6. Mix the small lumps of dough together.
7. Repeat with new colors.
8. Save the mixtures and let them dry.

## Discovery Questions

1. Compare color mixtures with peers—are all secondary colors the same?
2. What new colors were made?
3. Did anyone mix all three colors? What happened?
4. What would happen if you mixed unequal pieces?

## Integrated Curriculum and Extensions for Colorful Clay Unit

### Mathematics

Collect and count blue, yellow, and green objects in the classroom

### Social Studies

- skin color—the colors of friends in our world
- color of flags of countries

### Language Arts

**Literature**

Read Leo Lionni's *Little Blue, Little Yellow.*

**Vocabulary**

playdough • mix • knead • new • compare • separate

**Expressive language**

Retell the story *Little Blue, Little Yellow* in your own words.

**Receptive language**

With teacher's help, follow recipe to make playdough.


Chaille, C., & L. Britain. 1997. *The young child as scientist: A constructivist approach to early childhood science education.* New York: Longman.

French, L.A. 1996. "I told you all about it, so don't tell me you don't know": Two-year-olds and learning through language. *Young Children* 51 (2): 17–20.

Gallas, K. 1995. *Talking their way into science: Hearing children's questions and theories, responding with curricula.* New York: Teachers College Press.

Goodman, Y.M. 1986. Children coming to know literacy. In *Emergent literacy: Writing and reading,* eds. W. Teale, E. Sulzby, & M.Farr. Norwood, NJ: Ablex.

Katz, L., & D. McClellan. 1997. *Fostering children's social competence: The teacher's role.* Washington, DC: NAEYC.

Nelson, K. 1996. *Language in cognitive development: The emergence of the mediated mind.* New York: Cambridge University Press.

Teale, W., & E. Sulzby. 1986. Emergent literacy as a perspective for examining how young children become writers and readers. In *Emergent literacy: Writing and reading,* eds. W. Teale, E. Sulzby, & M. Farr. Norwood, NJ: Ablex.

# Entries from a Staff Developer's Journal . . .

# Helping Teachers Develop as Facilitators of Three- to Five-Year-Olds' Science Inquiry

Robin Friedrichs Moriarty

Rarely do we see three-, four-, or five-year-olds engaged in a science inquiry for months at a time, especially in classroom settings. Rarer still are materials that will support teachers as they learn to identify and deepen children's science understandings in the form of science inquiry. The National Science Foundation is helping make quality science inquiry in preschools, child care centers, and Head Start programs more common by supporting a curriculum development project at the Education Development Center (EDC) in Newton, Massachusetts, a nonprofit research and development institution.

EDC is writing three teacher guides, each with accompanying professional development materials, to help teachers of three-, four-, and five-year-olds identify science-rich questions embedded in their children's play and use those questions to engage their children in age-appropriate science inquiry. Two of the guides, "Exploring Water with Young Children" and "Building Structures with Young Children," focus on physical sciences, and the third, "Discovering Nature with Young Children," focuses on life sciences. Each guide describes ways teachers can prepare for and facilitate an inquiry that can last for eight weeks or longer.

For a month and a half, I supported three Head Start teacher teams as they used the "Exploring Water" materials for the first time. After each of our weekly two-hour sessions, I wrote in a journal and reflected on our work.

**Robin Moriarty** is a research associate at the Education Development Center (EDC) in Newton, Massachusetts.

The materials referenced in this article were authored by Ingrid Chalufour and Karen Worth for Education Development Center, with support from a National Science Foundation grant. Each is being published as a set (book, trainer's guide, training video) by Redleaf Press as *The Young Scientist Series:* "Discovering Nature with Young Children" (fall 2003), "Exploring Water with Young Children" (fall 2004), and "Building Structures with Young Children" (spring 2005).

Photos courtesy of the author.

## The journal discoveries

I knew from the start that I wanted to help teachers develop the confidence and the mechanical skills they need to facilitate a science inquiry over time. My journal entries describe afternoons spent studying the "Plan for Teacher to Follow" sections of the "Exploring Water" materials, rehearsing what teachers might do and say at each stage of the exploration.

But as I reread my journal entries, I saw something much more important. By connecting the basic skills they were learning to the reasons why those skills are key to their role as facilitators of inquiry, I helped teachers improve their ability to apply their new skills to future inquiries—inquiries that could be built around their children's unique interests and science-based questions, inquiries that would unfold without requiring the support of a teacher's guide. In short, as teachers incorporated the following three key principles into their practice, they developed the capacity to apply their evolving skills to new inquiries.

**1.** Teachers need a basic understanding of science, which involves both the understanding of science inquiry as a dynamic process and a familiarity with the science concepts they will help children explore. Therefore, before "Exploring Water" describes ways teachers can begin a science inquiry using questions that three-, four-, and five-year-olds typically wonder about, it leads teacher teams through an abbreviated science inquiry of their own, using the same water materials their children will be using later. This way, teachers feel prepared to act as guides to children, highlighting interesting possibilities and asking key questions as they focus children's thinking on the properties of water.

2. Children need access to thoughtfully selected materials that provide them with opportunities to work with and observe science phenomena. For example, when children have repeated opportunities to explore water with various-sized containers, eyedroppers, turkey basters, pumps, clear tubing, connectors, and funnels, they confront many of water's properties.

3. Reflection is an important part of the science inquiry process. Teachers need to provide various opportunities for children to revisit, represent, discuss, and demonstrate the experiences they have with these specially selected materials. The reflecting that children do on their explorations—for example, at the water table—not only helps children articulate and refine their theories about water flow but also gives teachers the information they need to focus the inquiry in ways that will help children examine their unique ideas about water flow more closely.

## Selected journal entries

The following excerpts from my journal as a technical assistant illustrate the ways six Head Start teachers gained important understandings about these key principles as they learned to use "Exploring Water."

### Excerpts from week one

Clarissa's question seemed like a good place to start: "I'm already doing science in my classroom. How is your approach different?"

"What does science look like in your classroom?" I wondered aloud. For the next half hour, we talked openly about our beliefs and values as they relate to science education for children from three to five years old. It felt good to get our biases out in the open. I heard about science tables, collections of pinecones, and units on the rain forest, and was handed a couple of well-loved science activity books.

Then it was my turn to describe my image of science for three- to five-year-olds, which is hard simply because inquiry is so dynamic. "For me, science inquiry begins with children's explorable questions," I explained. "When children ask, 'How can I get this water to move through these tubes?' we can guide them in an exploration to find out. But when they ask, 'Why does it rain?' we can't actively help them to explore. Why does it rain? is an intriguing question. It invites children to suggest explanations like 'Because the storm clouds come.' It can be fun to talk with children about their explanations for phenomena like rain and to find out more about their underlying ideas; you can also help them make observations, like noticing whether it really does rain every time storm clouds come. But the question of why it rains does not offer children opportunities to plan, conduct, and analyze a hands-on investigation."

Clarissa asked, "But if I don't understand the science of water flow, how can I help my children understand it?" We headed directly to the water tables! I always enjoy watching teachers explore water with clear tubing, turkey basters, funnels, pumps, and containers of different sizes and shapes, and today was no exception. When I positioned a piece of plastic-covered wire shelving down the middle of the water tables, they became especially intrigued and generated plenty of their own challenges.

Lucy and Clarissa worked to get water to flow over the shelving. Rose and Madeline attached funnels to each end of the shelf and found a way to direct water from one to the other. They named it a potion machine.

After we cleaned up, I charted the responses to my next science-focused question: What did you just notice about water flow? The responses included the following:

- water flows down;
- water will move up and down in a **U**-shaped tube if you move the sides up and down;
- pumps can make water flow faster; and
- bubbles of air sometimes get in the way of flowing water.

I summarized what I wanted the teachers to take away from this session: "Your children's understanding of water flow—that water usually flows down unless it is being pumped or squirted—will begin to develop as they openly explore different kinds of containers, tubing, and pumps with water, just the way your understanding did!"

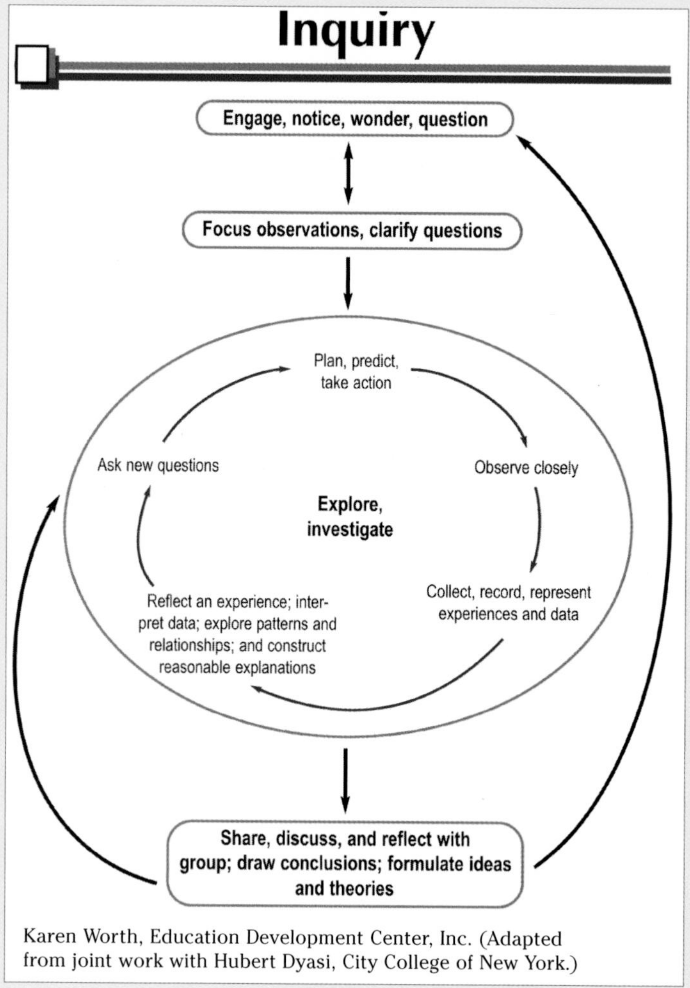

Karen Worth, Education Development Center, Inc. (Adapted from joint work with Hubert Dyasi, City College of New York.)

### Excerpts from week two

One week later, the teachers were still excited about having played with the water table materials! I'm relieved; I think it would be tough to move ahead if they were not excited. Today, I pulled out the inquiry diagram (see "Inquiry" at right), and we related last week's experiences at the water table to the graphic. (Each of the science explorations professional development packages includes this graphic of inquiry. It serves as a meta-cognitive tool that is best introduced after teachers explored the materials. In this way they can reflect on their own learning.)

One teacher described her experiences:

> At first I just wanted to pour water into every available object in the table! Then I focused on one of the wider tubes: I filled it, emptied it, filled it halfway and shaped it like a **U**, watched the water rise and fall as I moved the sides of the **U** up and down . . . now I see that I was moving from "wondering/exploring" to "taking action/extending" questions.

Yes! They saw the connection! But I needed to make it explicit: "And by engaging in those inquiry processes, you deepened your understanding of water flow!" It was time to move on. "I'm going to help you deepen your understanding of water flow even more!"

We returned to the water tables, but this time I took a more directive role. I wanted to move their inquiry into a more focused phase, so I challenged them to move water from the full table to the empty one—without carrying it! They connected tubes and funnels and pumped water into the funnel to push it through connected tubes and into the empty table. They were very proud!

As we reflected on their work, Sam connected these focused explorations to the inquiry diagram on the wall and to concepts of water flow written on a second chart. Clarissa wondered aloud, "What would happen if the full water table had no legs, and we had to move the water up and into the empty water table?" "Let's try it!" We cycled back into the inquiry! After referring to the inquiry diagram, we got to work.

### Excerpts from week three

We'd experienced some open and focused exploration of water flow as adults, but what about bringing water exploration to three-, four-, and five-year-olds? The power of the environment cannot be underestimated, and most teachers enjoy reorganizing their

classrooms and bringing in new materials, so this session turned out to be engaging.

"Having two water tables and other tubs of water around the classroom and outdoors says a lot about the importance of water exploration! And the water play materials are so inviting. I loved the clear tubing and the turkey basters."

Maxine wondered why we had not included dolls or boats in the water tables. "My children love both." We stopped to remind ourselves which science concept we were helping children develop at this point in the inquiry: water flow. Boats, we decided, would help bring the concepts of sink and float to the fore, but dolls? Nina stated it so perfectly: "Well, they're fun to wash, and they lend themselves to very early water play experiences, but they don't seem to add to a focused look at water flow."

"What about this?" I pulled out a couple of different-sized clear plastic cups that had a couple of holes punched in their sides. "What might these help children notice about water flow? Play with them and find out!"

Some of their discoveries included the following:
- I can make water squirt sideways without a turkey baster or a pump;
- streams of water shoot farther when the cups are full;
- the streams eventually turn into dribbles, and then they stop.

I couldn't help but state the obvious: "If I had given you dolls to wash instead of cups with holes to explore, you wouldn't have experienced these fascinating phenomena just now."

### Excerpts from week four

I worried that these teachers were going to think science inquiry with young children ends with a rich environment and time to pursue explorable questions. I wanted them to understand the important roles representation and reflection have in science inquiry. I struggled with how to help the teachers realize three-, four-, and five-year-olds' potential for reflection and processing. But using the overhead to share Daire's and Alessandro's drawings did just that.

"How can I get my children to create drawings like that!?" was the immediate reaction. We took the next half hour to discuss the mechanics of incorporating children's representational drawing into the classroom culture: expectations, materials, and procedures. But I didn't want to stop there. The next step was to practice analyzing representational drawings with two purposes in mind: to reflect on what they say about children's science understanding and to see how they suggest next steps for the inquiry.

First, we studied Daire's response to exploring water with these materials, which he expressed through his drawing of himself blowing into the longer tube. I asked

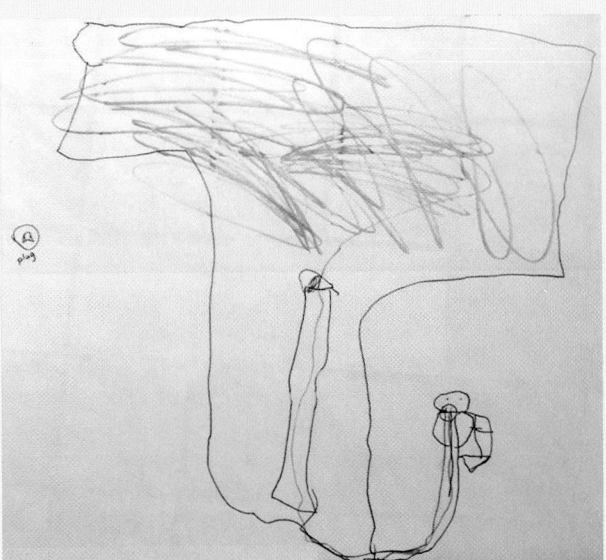

Daire's drawing

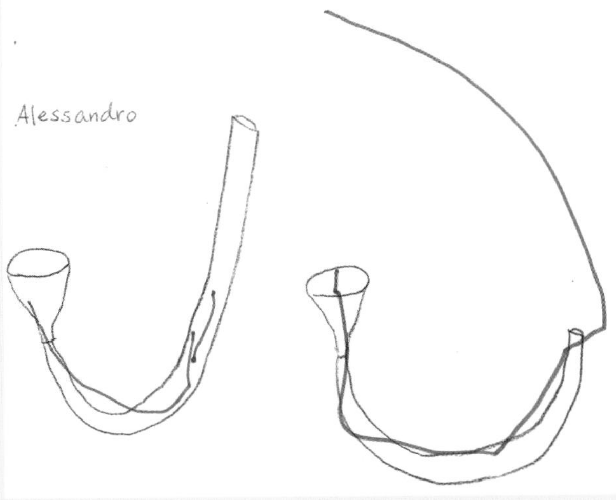


Alessandro's drawing

the teachers to consider what the drawing told them about his understanding of water flow.

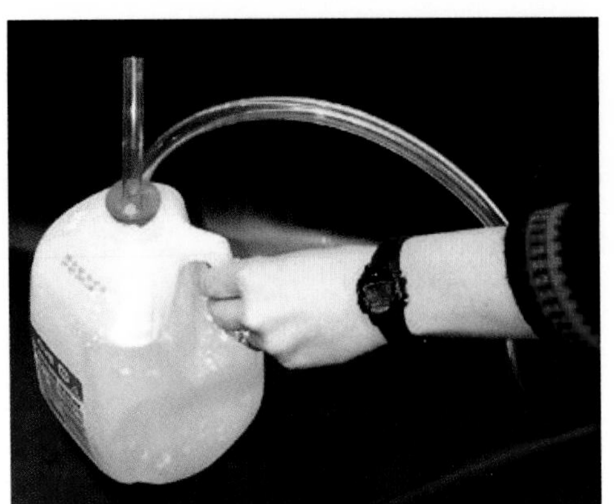

They concluded that Daire's drawing represents the experience he had making water spurt up and out of the shorter tube by blowing into the longer one. "Do you think he's developing an understanding of force's effects on water flow by noticing the way he can move water with his breath?" I asked. "Yes!" they declared. Next steps for him, they suggested, might include exploring water with different kinds of hand pumps.

The group decided Alessandro's blue marks on his teacher's black-line drawings represented the observations he made exploring water in a clear tube connected to a funnel. The drawing on the left, Clarissa suggested, represents his observation of water resting evenly, on the horizontal, and in the drawing on the right, what happened to the water when he raised the funnel. "Perhaps his observations will help him develop an understanding of gravity's effect on water flow," I suggested. "I think so!" Clarissa responded. Suggested next steps for Alessandro included adding **T** and **Y** connectors to his exploration of water in clear tubing.

Next, we practiced analyzing a third representational drawing by a child named Deon. Deon, I explained, used clear tubing attached with a **Y** connector to explore water flow, just the way we had. His representational drawing raised a question. The group agreed that it represents his observation of two water flows combining into one, but Rose and Carol wondered whether it also assumes the water will stop flowing when it gets to the bottom of the open-ended **Y** connector. Clarissa suggested that someone ask Deon to show what happened when the water he drew flowed into the **Y** connector. Her suggestion gave me the opportunity I was looking for to stress the effect children's representation has on deepening science inquiry: First, I explained, it gives children an opportunity to revisit their explorations, and second, it presents them with repeated opportunities to reflect on their work and the work of their peers throughout the inquiry.

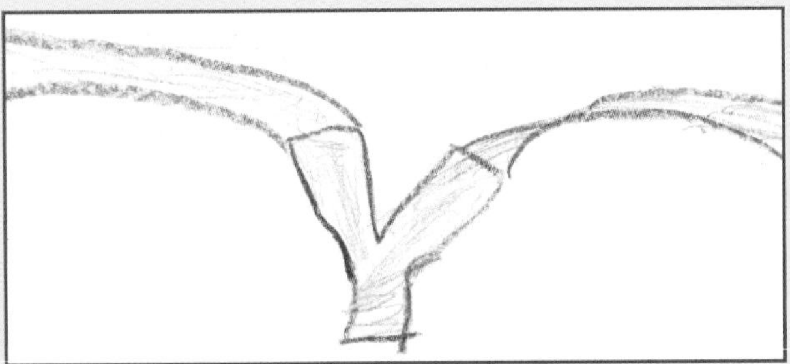

Deon's representation

I suspect we will use our last session to focus quite a bit on strategies they can use to facilitate science-focused conversations with children, whether they be initiated by a piece of representation, a book, a demonstration, or a photograph. There are so many ways to incorporate reflection into science inquiry!

## Postscript

I ran into Clarissa today. She is so excited! Some of her children are fascinated with her overhead projector. The other day, she observed a group of four covering the top of the projector with pattern blocks. When they turned the projector on, they were surprised to notice that their colorful pattern wasn't shining on the wall!

Clarissa decided these children have questions about light and shadows, so she borrowed a light table and explored the way different materials looked on it. She also made her own shadow puppets and went on shadow walks in- and outdoors. I asked her what concepts she plans to keep at the center of the inquiry, and she explained that she hopes to help children think about shadows: how they're made, how different materials respond to being lit up from behind or from the front. She is already using two of the key principles to help her develop this inquiry: She has got a basic understanding of the science concepts at the core of the inquiry, and she has collected some intriguing materials to help children explore those concepts.

Clarissa has invited me to meet with her next week so we can brainstorm ways she might help her children represent and share their exploration of light, shadows, and different materials. She has remembered the importance of helping children reflect! But reflection is also key to developing teacher practice; I've decided to make her a blank "Light and Shadows Inquiry Journal" so I can give it to her when we meet.

# Using Photographs to Support Children's Science Inquiry

Cynthia Hoisington

As a Head Start teacher in a full-day, inclusive classroom, I often take photographs of the children in my class and use them in a variety of traditional ways. Recently, however, I discovered a new use for photography, as a powerful teaching tool to support science learning.

It all began when my class and I were invited to participate in the Education Development Center (EDC) science curriculum development project "Building Structures with Young Children." My role as a development teacher in the project was to observe, support, and extend children's building play; assess what science-related questions they were asking themselves; and help children frame and organize their building experiences by supporting building-related discussion and representation.

To document what children were doing and to share what was happening in my classroom with other members of the curriculum development team, I decided to capture the action with a digital camera. I envisioned being able to use the photos as tools for both assessment and planning. Examining photographs of the children building structures would let me reflect on what the children knew and how they were using the construction materials. In turn, I could plan further activities and experiences, building on children's interests and current knowledge (Helm, Beneke, & Steinheimer 1998).

I knew that photographs would also be useful in telling children's families and my colleagues the story of what we had learned during the building unit. I planned on collecting the photos I would take throughout the course of the unit, captioning them with children's descriptions of what they were doing, and arranging these photos and descriptions on poster board along with children's building-related drawings. I would add my own commentary to the structures display, focusing on how the children's explorations supported science, math, and literacy learning.

However, as the children and I became more and more immersed in the Building Structures Exploration project, several new, unplanned uses of photography emerged: to help children revisit and extend their investigations, to reflect on their building experiences and articulate the strategies employed, and to analyze and synthesize these building strategies. These uses turned out to be the most exciting ones of all.

## Photographs help children revisit and extend their investigations

From the very start the children were excited about building, but something was missing. Children would build one thing one day, something different the next; they didn't seem to be making any connections between their various building experiences.

One day I took a photograph of Christine next to several towers of cylindrical unit blocks she had made. As I looked at the photo later, I got the idea to use it as a basis for a class discussion. The next day during circle time I shared this photo with the whole group and invited Christine to tell us what she was doing in the picture. She responded, "Building with fat round blocks and skinny round blocks!"

**Cynthia Hoisington,** B.S., is director of child development services at the A.B.C.D. South Side Head Start program in Boston. Cynthia is currently a graduate student at Bridgewater State College in Massachusetts. For the past three years she has worked with curriculum developers at the Education Development Center in Newton, Massachusetts, on the science curriculum. The materials referenced in this article are being published by Redleaf Press, under the title *The Young Scientist Series.*

Photos courtesy of the author.

Together as a group we counted the number of blocks in each of Christine's towers, and I introduced the words *taller, tallest, shorter,* and *shortest.* When it was time to transition to free-play time, I noticed that Christine was anxiously looking at the unit blocks. At her turn to choose an activity area, she said excitedly, "I want to build with fat round blocks again!" She immediately began to recreate her tower of the day before. Showing Christine the photo of herself building, and inviting her to share it with the group, was enough to inspire her to persist in her building.

The following day Christine took out one of the hard plastic boards I had placed in the block area and used it as a base. I watched as she quickly placed the blocks, one on top of another. I resisted the urge to give her advice about how to place them. When the new tower reached the same height as the tallest original one, I asked Christine if she wanted to try going higher. She seemed nervous but excited as she carefully placed an additional block, using both hands. "I notice you are placing that block very carefully," I said. "I don't want it to fall down!" she answered.

Our group discussion with the photo had clearly inspired her to continue the activity. In returning to the same type of building, she was able to confront the challenges of design, balance, and stability that she had encountered the day before, but this time she tried new strategies.

As the children became accustomed to having their building experiences photographed, they took a more

**Several new, unplanned uses of photography emerged: to help children revisit and extend their investigations, to reflect on their building experiences and articulate the strategies employed, and to analyze and synthesize these building strategies.**

active role in documentation. I began to rely on them to tell me when they wanted pictures taken of specific structures. They also asked about taking their own photos. I eventually got an Instamatic camera the children could use, although I provided it only when children were clearly engaged in a building activity and when I saw that taking photos of their structures would help them to think more deeply about what they were doing (film is expensive too). I encouraged the children to think about what they wanted to show in their photos before they began snapping pictures.

### Photographs help children reflect on building experiences and articulate their strategies

After several weeks of building, many of the children became interested in building high towers. But I noticed that they often chose blocks haphazardly, without making any connection between what they wanted to build and what blocks they chose to use. I also noticed that younger children and less experienced builders seldom made a connection between their own building behaviors and the resulting structure.

With photographs, I was able to help Ray-Shawn and John overcome the disconnect problem. Day after day, the two boys had been piling up cardboard blocks without seeming to think about how they could increase the chances of their tower's staying up. In an effort to encourage their thinking, I asked, "How do you think you could get the building to stay up this time?"

"I know! We can wear hard hats," Ray-Shawn said as he took two off the shelf for himself and his friend. With the hard hats on their heads, they stacked the blocks up quickly once again, and again the structure toppled. "The hats don't work," said Ray-Shawn sadly.

I wanted to help the children focus on their own building strategies, as well as the characteristics of the blocks, and consider how these things affected the structures they were making. To do this, I created a challenge for the children and photographed the results. I asked them to use four different types of blocks to answer the question, "Which blocks will make the tallest tower?" Then I brought to the group four selected photographs from this activity, and we used these photos to compare the towers in different ways.

We first counted the blocks in each structure and next checked to see how high up each tower came on the child standing next to it in the photo. Nora's foam-squares tower only came to her knee, but Christine's unit-block tower was much taller than she was and

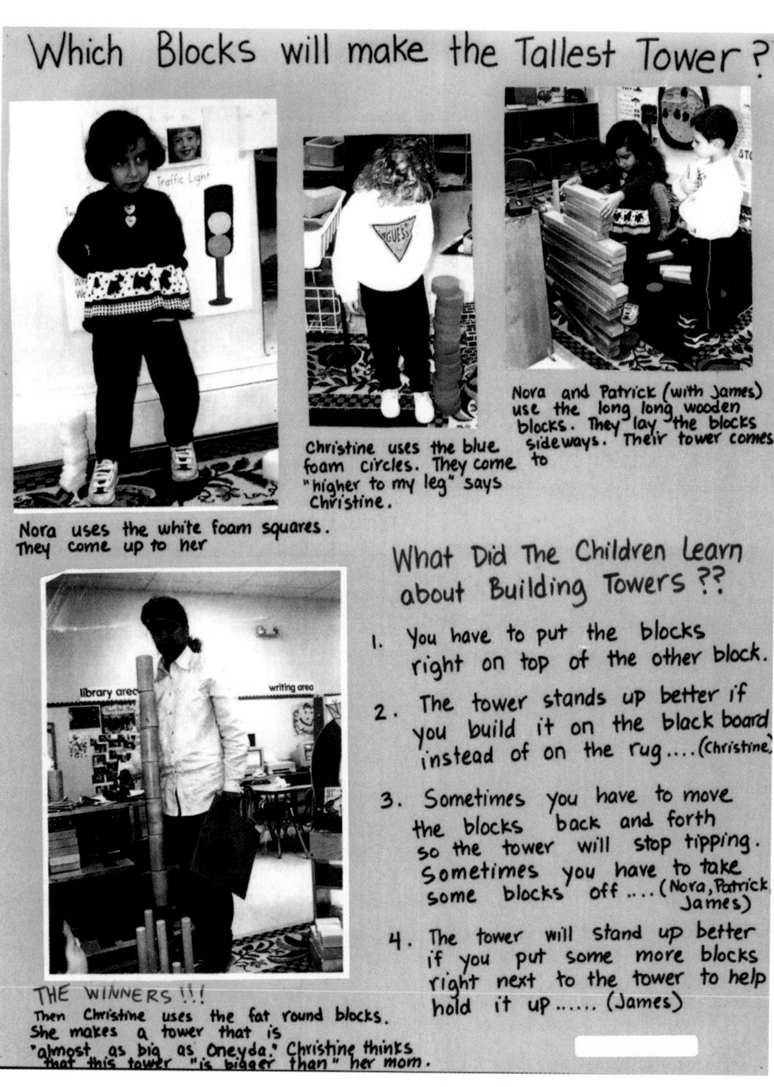

Which Blocks will make the Tallest Tower?
Nora uses the white foam squares. They come up to her
Christine uses the blue foam circles. They come "higher to my leg" says Christine.
Nora and Patrick (with James) use the long long wooden blocks. They lay the blocks sideways. Their tower comes to
What Did The Children Learn about Building Towers ??
1. You have to put the blocks right on top of the other block.
2. The tower stands up better if you build it on the black board instead of on the rug....(Christine)
3. Sometimes you have to move the blocks back and forth so the tower will stop tipping. Sometimes you have to take some blocks off....(Nora, Patrick James)
4. The tower will stand up better if you put some more blocks right next to the tower to help hold it up...... (James)
THE WINNERS!!!
Then Christine uses the fat round blocks. She makes a tower that is "almost as big as Oneyda." Christine thinks that this tower "is bigger than" her mom.
library area
writing area

## Pluses for Digital Cameras in the Classroom

During the "Building Structures" project, I used a digital camera. I found it had several advantages over the standard cameras I had used in the past. Digital cameras can be quite inexpensive.

**1. The miniviewer allows the teacher and children to see the photo directly after it is taken.** This feature made it possible for us to discuss photos immediately. Also, with a digital camera I could discard poor-quality photos without the expense of printing them first.

**2. Editing photos on the computer allows the teacher to emphasize specific building materials and the strategies she wants children to focus on.** For example, as I looked at the photos of Kamela and Christina's houseboat, I realized I wanted them to look more closely at exactly how they were placing individual waffle blocks together. Computer editing let me zoom in on their hands placing the blocks and enlarge these portions of the photo.

**3. The teacher can manipulate the size of the prints.** It was much easier for the children and me to look at structures, especially to observe design details, when I made larger prints.

**4. Saving the photos to a computer file allows access to all or any of the photos taken during the unit at any time.** Being able to look at all the photos taken over a period of time helped me to plan further explorations. For example, after reviewing the photos of children building with blocks that were taken over a period of weeks, I realized they were especially interested in learning how to build tall structures.

**5. Individualizing is easier.** Since I could also save photos of specific children in their own files, it was possible to observe how each child's building activity changed over time and to work with a child individually.

clearly the highest of the four. "Fat round blocks make the biggest towers!" Christine exclaimed happily.

"Why do the fat round wood blocks make the tallest tower?" I wondered out loud. "What do you think?" I asked the children. "Because those blocks are bigger," said Christine. "Because they are harder," said Ena. "Because they are heavy," said Patrick.

Photos helped the builders reflect on aspects of their buildings. Sharing the photos helped other children solve design problems they were experiencing. One day Dayvian and Patrick were trying to make a roof on their waffle-block house, but it kept caving in. When we looked at photos of the house the next day, I asked Patrick what he was trying to do. He said, "Put a roof so the rain don't come in the house." I asked him to describe the problem they were having: "The roof keep falling down." Another child piped up, "It breaking in the middle."

As Patrick looked more closely at the photo he said, "The roof too big! Come on, Dayvian. We gots to make it smaller!" This conversation prompted Patrick and Dayvian to make the roof in a new way. Instead of making it two blocks wide they made it one block wide, thus avoiding the whole problem of a buckling seam in the middle.

### Photographs help children analyze and synthesize data

I first realized that photos could be used as tools for analysis during a particular building sequence in which Kamela and Christina were engaged. Over a period of days they tried to build a houseboat using large waffle blocks. As I observed them, I realized they were using a hit-or-miss process. They had not discovered that each waffle block had to be fitted to the next in a specific way for additional blocks to fit correctly.

I began to take photos that focused specifically on their hands as they put the edges of individual blocks together. When I shared these photos with them, we talked about how they were fitting together the slots in adjacent blocks. Then I asked them if there was a certain way that always worked.

When Kamela and Christina went back to work on their houseboat after our discussion of the photos, I noticed a change. Now when they attached two blocks, they immediately checked to see if the blocks aligned

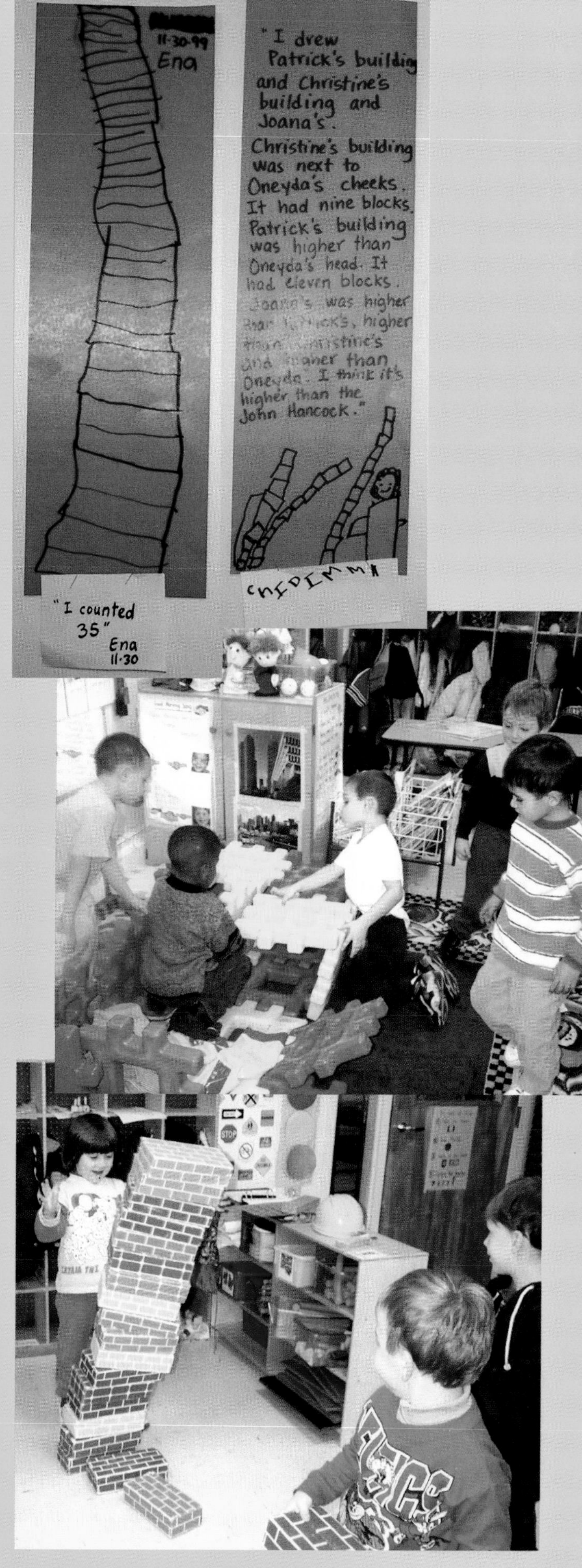


**Photos helped the builders reflect on aspects of their buildings. Sharing the photos helped other children solve design problems they were experiencing.**

correctly before adding a third block. In studying the photos and talking about their methods, they had begun to understand there was a pattern to the way the blocks had to be attached.

This experience made me think about how I could help all of the children to see patterns and make generalizations about what they had learned. I started by focusing on tower building strategies since towers had been the most popular activity. At circle time I shared photos with children that showed the strategies they had developed and used to build high towers, including placing blocks carefully, building on a hard surface, and placing other blocks as supports next to their buildings.

As we looked at each photo, I pointed out the strategy I saw the child using and asked the child to talk about it. I said to James, "I notice you put another pile of blocks next to your tower. What were you thinking about when you did that?" He said, "I put more blocks together to my tower so it don't fall down!"

I continued, "James said these blocks next to his building help it to not fall down. Does anyone have any ideas about how that works?" One of the five-year-olds speculated, "Because it pushes it up." "Oh," I said, "you think the blocks next to James's building push it up like a brace for his building? Maybe we can look for braces on structures outside when we go for a walk today."

When I asked Christine why she decided to build her tower on a hard plastic board, she said, "Because the rug is too softer!" Another child said, "The rug is slippery. It makes the blocks fall." "Hmm," I said, "the rug is soft and slippery? And what about the hard plastic? Why do you think that works better?" "Because the blocks stick better, they don't slip," said another five-year-old.

I had noticed that children sometimes used the idea of "sticky" when they seemed to be thinking about balance. "Do you think the blocks can balance better on

**In studying the photos and talking about their methods, the children had begun to understand there was a pattern to the way the blocks had to be attached.**

the hard board?" I asked. "Yeah," the children nodded. Once we had generated a list of the particular strategies children used to make towers, I posted it on a documentation panel along with photographs that illustrated the towers challenge.

As tower building continued, children's unit-block towers continued to get higher, and I continued to take photos. I used these photos, along with the original photos and our list of strategies, to draw their attention to the fact that they were using these same techniques over and over again.

I knew that helping children to generalize about why certain materials and certain strategies were more successful for making strong buildings was a gradual process. But I could see that the children were beginning to think about it. They were beginning to see patterns and make connections between their own explorations and discoveries and what we observed outside in the world. I especially knew that using photos had been instrumental in helping them to do this.

## Picture-Taking Strategies for Educators

- Keep the camera within easy reach in the classroom and loaded with film (or charged if using a digital) at all times.
- Take photos unobtrusively, yet often enough so that children get used to seeing you with a camera in your hand.
- Capture the entire structure when photographing any structure.
- Take individual photos of various structures from approximately the same angles and distances for easier comparison later.
- Have a goal in mind when you take photos of children building. Your goal will influence what you decide to focus on.
- Be spontaneous sometimes too. If something that children are building draws your attention or piques your interest, snap it. Don't worry if you don't have a goal for every photo.

## Conclusion

My experience using photographs with children during the "Building Structures" project was a powerful one for me as a teacher. It taught me that photographs are useful tools not only for literacy learning, social-emotional development, and assessment but also for science teaching and learning.

During the children's building exploration, photographs served as bridges between their various single building experiences and helped them to persist in building. Photos provided the means for children to step back from their actual building experiences so they could reflect and think specifically about materials and strategies. Using the photos as props for discussion allowed children to share their experiences, to hear other perspectives, and to generate new ideas for solving building problems.

The photos also supported children in their developing abilities to represent their building experiences symbolically and in encountering new challenges in the process. Photographs of multiple building experiences, taken over a long period of time, helped children to analyze what they had learned about the materials and specific building strategies and to make generalizations about their experiences.

As a preschool teacher who wants to support science learning, the camera has now become a routine piece of equipment in my classroom.

## Reference

Helm, J.H., S. Beneke, & K. Steinheimer. 1998. *Windows on learning: Documenting young children's work.* New York: Teachers College Press.

# Documenting Early Science Learning

Young children are fascinated by the natural world. They explore how things work, wonder what is and is not alive, and think about why some things change their shape and form. Science explorations such as planting seeds, studying animals, and baking bread are a natural part of early childhood classrooms and can provide the settings in which teachers can observe how children are making sense of the world around them. The real evidence of children's early science understanding comes directly from these everyday experiences.

Records of children's conversations, anecdotal notes and photographs of their actions, and samples of their drawings and constructions all form the classroom-based data that helps teachers learn how children are thinking about the natural world. The documentation process itself helps teachers gain a deeper understanding of individual children in the class *and* enhances general knowledge of how young children make sense of the world. Finally, by engaging in the documentation/assessment process of collecting, describing, and interpreting evidence of young children's emerging science understandings, early childhood educators are able to provide more appropriate science-related experiences and learning environments.



**Jacqueline Jones and Rosalea Courtney**

## Guiding principles

Three principles developed through a series of ongoing collaborations between educational researchers and preschool and early elementary teachers guide the classroom-based documentation process (Chittenden & Jones 1998). The principles reflect sound practice in educational measurement across the developmental continuum and across content areas (Shepard, Kagan, & Wurtz 1998; American Educational Research Association, American Psychological Association, & National Council on Measurement in Education 1999).

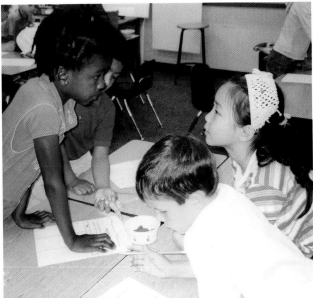

**Jacqueline Jones,** Ph.D., is a senior research scientist in the Research and Development Division at Educational Testing Service in Princeton, New Jersey. Jacqueline studies assessment in early childhood education, specifically classroom-based strategies to document young children's science and literacy learning.

**Rosalea Courtney,** M.A., is a senior research associate in the Teaching and Learning Research Center at Educational Testing Service. Rosalea has been involved in a number of preschool and elementary education research projects. She has worked closely with teachers to investigate ways to document and assess children's science and literacy learning.

This material is based on work supported by the National Science Foundation,Grant No. 9731282.

**The real evidence of children's early science understanding comes directly from everyday experiences.**

### 1. Collect a variety of forms of evidence

It is important to collect a variety of records because children vary in how they convey their ideas. Some children demonstrate their understanding through constructions or drawings while others are more comfortable talking about what they see or think. Educators can learn a great deal about children's thinking by listening carefully to their language *and* looking at samples of their work. Examples of various forms of evidence that are a part of most early childhood classrooms include the following:

***A drawing***—A preschool child's observations of the spots on the class rabbit, the shape of its ears, and its bushy tail can be seen in her drawing.

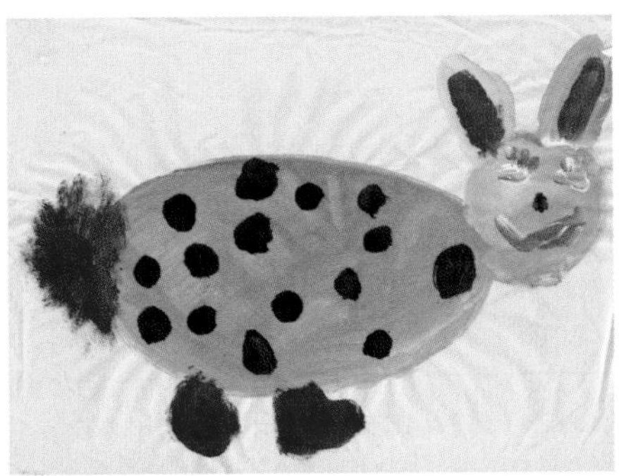

***Drawing and dictation***—Drawings alone may not always reflect a young child's ideas and perceptions. The child's thinking about the rabbit (named Baby) is more visible when the child's dictated comments are added to the drawing.

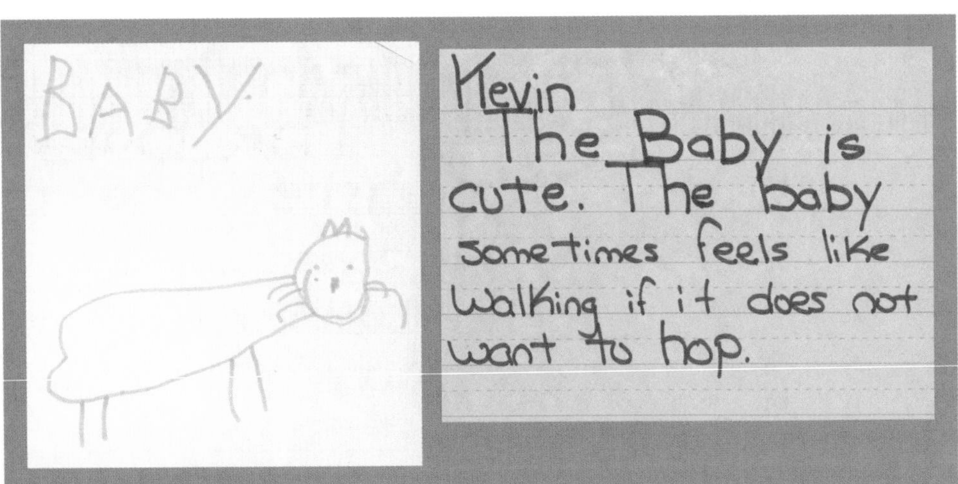


***Photographs***—Images of children at play can reveal their emerging science thinking. After a preschool class had finished shucking corn, the teacher photographed this child's spontaneous construction of a "corn garden" in the block area. She pretends to water the garden.

© Rick Toone, ETS

***A record of children's language***—Making records of children's conversations can reveal children's emerging ideas (see "How Does the Moon Change?" on p. 29). Collecting children's language works best when teachers use open-ended questions that invite the children's participation. Questions such as What have you noticed about . . . ? or What might happen if . . . ? or What are some things you know about . . . ? imply no single best answer. Instead, they encourage children to respond from their own observations, experiences, or conjectures.

### 2. Collect the forms of evidence over a period of time

Classroom-based evidence should be collected over a period of time because young children's learning is not linear. Rather, it is episodic and based on individual experiences. Any single piece of evidence captures just one moment in time when a child may be struggling with an idea or question. A teacher who collects evidence over a period of time can see the evolution of an idea or concept. For example, a review of entries in children's science journals over several weeks can reveal the development of more focused and detailed observations.

## How Does the Moon Change?

**GRADE K/1 CONVERSATION**

***Gale:*** The moon doesn't really get smaller and smaller. It just changes its shape.

***Tommy:*** It DOES get smaller and smaller. That's the only way it can change its shape.

***James:*** It gets smaller and it goes behind the clouds and changes its shape, or there's another moon that's another shape.

***David:*** Everyone thinks the moon is low but it really is high but it gets lower one day at a time.

***Gale:*** Sometimes when you're walking at night, it looks like it's following you but that's because you're moving.

***Steven:*** If you're walking, the moon is walking with you.

***David:*** It seems like the moon is moving when you're walking because the whole world is moving slowly and the moon is moving too.

***Steven:*** When you're driving in a car, everyone's going slowly but the moon is going fast.

***Sam:*** When you're riding in a car, the moon is following the car. The car is fast and the moon is too.

***Jada:*** Every car has a thin string that pulls the moon.

Children's understandings of big ideas such as life processes and changes in matter are not established firmly with one experience. Children need time to return to these ideas and concepts, to ask new questions, and to fit new learning into established ideas. The evidence of young children's learning is most useful when it is viewed over a period of weeks or months.

For example, some children in a first-grade classroom observed the life cycle of silkworms from eggs to adult moths over several weeks. The children wrote their observations in their science journals (see "Life Cycle of Silkworms: Science Journal Entries" on p. 30). In one child's journal, a number of entries describe the silkworms' size and color: "They are brown when they are little and they are green and brown when they are big." She also notes differences between the newly hatched caterpillars and human babies: "The little ones that just hatched is not moving like us when we were babies. We did not know how to do anything like them." In a later entry she writes about how the silkworm caterpillars are dependent upon humans to find food for them: "They don't live in trees. They don't live in the ground like the other worms. They don't live by themselves. You have to take care of them."

### 3. Collect evidence on the understanding of groups of children as well as individuals

Science is an inherently social activity, and children should be encouraged to discuss their ideas with one another. Collecting evidence of group learning helps the teacher to get the bigger picture of what the class as a whole is questioning or coming to understand about a concept. In addition, group evidence can give the teacher a better sense of what children bring to a topic, what they share, and where there are experiential differences. When evidence is collected for groups of children, patterns in thinking become apparent. For example, group conversations at the beginning of a unit can identify prior knowledge and can highlight misconceptions that are shared.

In one case, a kindergarten class had been observing caterpillars as part of their study of the life cycle of butterflies. During circle time, the teacher asked the class,

## Life Cycle of Silkworms: Science Journal Entries

**4/10:** They are brown when they are little and they are green and brown when they are big.

**4/18:** Their whole body looks like a elephant's trunk.

**4/25:** The little ones that just hatched is not moving like us when we were babies.

**4/26:** Their heads is fatter than their body. And their heads are white and their body is brown.

**5/1:** Now they are growing and changing colors. When they are babies, they were brown, but now they are white.

**5/2:** You have to take care of them. You have to give them some leaves. One is eating and Tyler is right. They do start by eating from the edge.

**5/4:** Now they are growing and they are learning how to eat fast. But when they were babies, they eat slow, but they are not little any more so now they eat faster. They were very hungry and my big fat one is eating fast just the way I said . . .

**5/10:** I have a question. What's the black stuff on top of their face?

**5/11:** One of my silkworms were making silk and I thought it was shedding skin, but it was not shedding skin. It was making silk.

**5/19:** When I cleaned out my container and gave my silkworms some leaves . . . some of them have to share and they was fighting together and then was eating and one silkworm eat a whole leaf. And he still eating another leaf. He was really hungry.

"What do you think is happening to these caterpillars?" Each child had an opportunity to respond to the question. The teacher wrote out each child's statement, producing a record of the class discussion. (See "What's Happening to These Caterpillars?")

In another example, preschool children made this collection of drawings when a rabbit was brought to class. Although there were no carrots in the classroom during this visit, each of these drawings shows the class rabbit with the food. The association between rabbits and carrots appeared to be shared by many of the children.

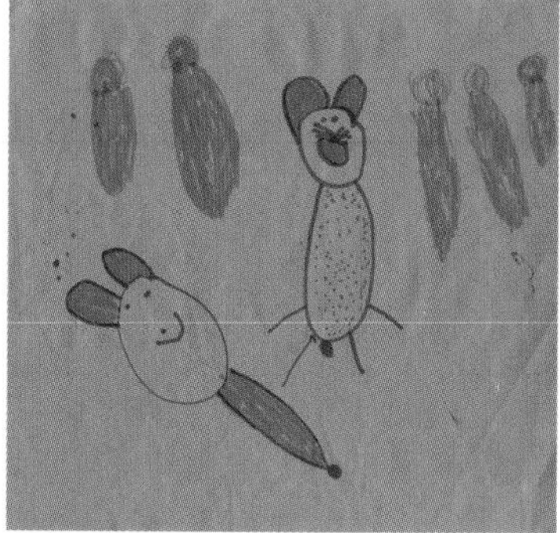

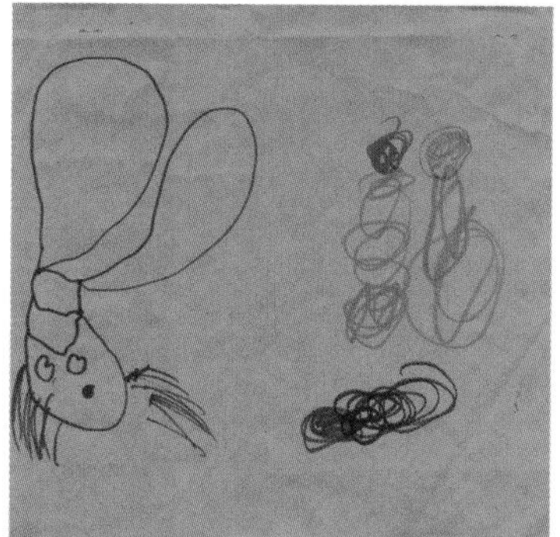

## What's Happening to These Caterpillars?

***Jason:*** They don't look like caterpillars. They look like celery 'cause they are green.

***Gabrielle:*** Two of them are starting to change into cocoons. They look like celery. One of the caterpillars is moving.

***Morris:*** One is already in a cocoon. Four are still caterpillars. One is not moving so it's a cocoon. Cocoons can't move.

***Sam:*** I see one moving.

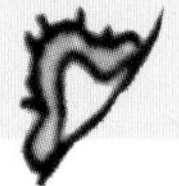

***Jaella:*** One has something sticking out.

***Kiri:*** Two look like trees. Some stuff is sticking out of the body.

***Tom:*** One looks like a leaf. One is all squished up.

***Zack:*** Two don't look the same. One looks like it has bumps on the side.

***Jason:*** They don't look like themselves so they must be cocoons. They're not long like the others.

## The documentation and assessment process

With the guiding principles as a foundation, documentation and assessment of young children's emerging science understandings consist of a five-stage cycle of identifying, collecting, describing, interpreting, and applying the classroom-based evidence in order to plan more appropriate experiences and environments.

**• Identify appropriate science-related goals and concepts, activities and experiences, and classroom settings.**

It is important to have some agreed-upon notion of what we as educators want children to experience, explore, and understand. In addition to specific curriculum goals, teachers who participated in this documentation process often used the *Benchmarks for Science Literacy* (American Association for the Advancement of Science 1993) or the *National Science Education Standards* (National Research Council 1996) to guide their expectations for young children. These documents were especially useful in providing a focus for collecting those samples of children's work that highlight specific science goals.

**• Collect evidence of children's learning, including records of children's conversations and children's work samples.**

Records of children's conversations and their work samples can take a variety of forms, including whole-class discussions, individual interviews, drawings, constructions, and diagrams. Consider which forms of evidence will give the best indication of how children are coming to understand the selected science goals and concepts. For example, many teachers have found that asking a child to dictate a description of his drawing and attaching the child's comments to the drawing can provide more information about the child's understanding than the drawing alone.

**• Describe evidence of children's learning without judgment and discuss it with colleagues.**

The first step in understanding what children are learning is to take a close look at what is actually in their language records, drawings, and constructions before reaching a conclusion (Himley & Carini 2000). Description is a skill that takes some practice. It may be easier to see what is missing or incorrect in children's statements or work samples than it is to focus on the knowledge that is represented. Working with colleagues on a careful description of children's work samples and records of their language can provide new and useful insights into children's learning (see "Steps in Guiding a Discussion of Children's Language Records"). Another teacher, or a parent, can bring a new perspective, often seeing things in the work that the child's teacher may miss.

### Steps in Guiding a Discussion of Children's Language Records

1. The teacher presenting the language records provides contextual information for the group: the science-related goal(s), science topic, classroom setting and activity, and specific prompt (question or set of directions).
2. Colleagues read the language record silently, or participants take turns reading a line of the record aloud.
3. The presenting teacher and colleagues discuss the record using descriptive statements *only*.
4. The group summarizes the descriptive statements.
5. Colleagues ask any additional questions related to the context, the child, or the record for clarification.
6. The group identifies how statements and questions in the language record reflect the science learning goals for the class.

**• Interpret evidence of individual and group understanding by connecting to learning goals and identifying patterns of learning.**

At this stage the children's work should be compared to the standards and goals identified by the teacher at the start of the cycle (Stearns & Courtney 2000). Does the work demonstrate the intended goals, such as observation or prediction? Are some additional types of work samples needed to demonstrate understanding? Are patterns of understanding emerging for the whole class? For example, if the teacher wants the children to observe living things in the classroom—noting changes that take place over time—and

**Group evidence can give the teacher a better sense of what the children bring to a topic, what they share, and where there are experiential differences.**

to ask questions about their observations, then journal entries can be used as evidence that these goals have been met.

• **Apply new information and understanding to improve instruction and curriculum and future assessment.**

The major purpose of assessment is to inform instructional practice. Therefore, the information from the documentation and assessment process must be tied directly to new planning. The process begins anew as the teacher uses information and insights gained from the process to identify the next set of the science-related goals and experiences. The cycle continues with children's emerging science understandings being nurtured and documented in the everyday life of an early childhood classroom. Teachers have found this process valuable for understanding the learning of individuals and groups, for guiding instruction, and for reporting to parents.

## References


American Association for the Advancement of Science (AAAS). 1993. *Benchmarks for science literacy.* New York: Oxford University Press.

American Educational Research Association, American Psychological Association, & National Council on Measurement in Education. 1999. *Standards for educational and psychological testing.* Washington, DC: American Educational Research Association.

Chittenden, E., & J. Jones. 1998. Science assessment in early childhood programs. In *Dialogue on early childhood science, mathematics, and technology education.* Washington, DC: American Association for the Advancement of Science.

Himley, M., & P.F. Carini, eds. 2000. *From another angle: Children's strengths and school standards. The Prospect Center's descriptive review of the child.* New York: Teachers College Press.

National Research Council. 1996. *National science education standards: Observe, interact, change, learn.* Washington, DC: National Academy Press.

Shepard, L., S.L. Kagan, & E. Wurtz. 1998. *Principles and recommendations for early childhood assessments.* Washington: National Education Goals Panel.

Stearns, C., & R. Courtney. 2000. Designing assessment with the standards. *Science and Children* 37 (4): 51–55, 65.


## National Science Education Standards

Excerpt from Content Standards: K-4 (http://www.nap.edu/html/nses/html/6c.html)

### Science as Inquiry

*Content Standard A: As a result of activities in grades K-4, all students should develop*

**abilities necessary to do scientific inquiry.**

*Ask a question about objects, organisms, and events in the environment.*

This aspect of the standard emphasizes students asking questions that they can answer with scientific knowledge, combined with their own observations. Students should answer their questions by seeking information from reliable sources of scientific information and from their own observations and investigations.

*Plan and conduct a simple investigation.*

In the earliest years, investigations are largely based on systematic observations. As students develop, they may design and conduct simple experiments to answer questions. The idea of a fair test is possible for many students to consider by fourth grade.

*Employ simple equipment and tools to gather data and extend the senses.*

In early years, students develop simple skills, such as how to observe, measure, cut, connect, switch, turn on and off, pour, hold, tie, and hook. Beginning with simple instruments, students can use rulers to measure the length, height, and depth of objects and materials; thermometers to measure temperature; watches to measure time; beam balances and spring scales to measure weight and force; magnifiers to observe objects and organisms; and microscopes to observe the finer details of plants, animals, rocks, and other materials. Children also develop skills in the use of computers and calculators for conducting investigations.

*Use data to construct a reasonable explanation.*

This aspect of the standard emphasizes the students' thinking as they use data to formulate explanations. Even at the earliest grade levels, students should learn what constitutes evidence and judge the merits or strength of the data and information that will be used to make explanations. After students propose an explanation, they will appeal to the knowledge and evidence they obtained to support their explanations. Students should check their explanations against scientific knowledge, experiences, and observations of others.

*Communicate investigations and explanations*

Students should begin developing the abilities to communicate, critique, and analyze their work and the work of other students. This communication might be spoken or drawn as well as written.

You can read all of the National Science Education Standards online at **http://www.nap.edu/html/nses/html**

# Be a Bee

## and Other Approaches to Introducing Young Children to Entomology

James A. Danoff-Burg

Although we human beings like to think we are the most important creatures on earth, statistically, we are the minority of its inhabitants. Insects lead the count, comprising approximately 53 percent of animal life. And just think of this: 20 percent of all known living species are beetles!

Clearly, it is in our interest to be comfortable with the joint-legged insect neighbors who outnumber us. Yet many adults regard insects with caution, even fear. The creatures have been universally typecast as unpleasant, even dangerous. We feel revulsion when they get into our homes and food. They anger us when they sting us in self-defense or, worse yet, feed on us. In truth, only a small minority of insects—less than one percent—pose any kind of problem for humans (Wilson 1992). Most species avoid contact with people and the environments that we create. Therefore, the aversion that many adults feel for insects is unfounded.

Perhaps our best bet for changing attitudes is to begin early. Most children—unlike adults—are fascinated by insects. The primary grades are an especially fertile time for nurturing an interest in insects. Many primary school–age children, who are learning about the world, are captivated by bees' communication skills and the caterpillar's ability to grow by shedding its skin or to change into a butterfly.

If we capitalize on children's inherent interest in insects and sustain this interest before it changes into a fear of nature, we can engender a lifelong respect for insects. Teachers can encourage children's interest by incorporating the study of insects—known as *entomology*—into the curriculum.

> **Children need opportunities to explore insects in depth, investigate insects' habits, share observations, and internalize what they have learned.**

For many children, their first encounter with insect knowledge is through a museum display of stunning—but dead—insects pinned to a burlap-covered board or in a case. Museum staff commonly refer to these exhibits as Oh-my! collections after the oft-uttered reaction of children. However, while most children find these collections initially intriguing and ask such questions as why the wings are so big or why a beetle has such large forelegs, they ultimately become bored and uninterested because the displays are passive and undemanding. Activities that invite children's active involvement are clearly needed (Lind 1999).

To promote interest and truly effect attitude change, an introduction to entomology should be more than a one-shot experience with dead animals. Children need ongoing exposure to and exploration of insects, beyond what any single visit to a natural history museum can provide. They need opportunities to explore insects in depth, investigate insects' habits, share observations, and internalize what they have learned. Above all, they need time to engage in active learning.

---

**James A. Danoff-Burg,** Ph.D., is a research scientist interested in habitat fragmentation and invasive insects. He is the director of Columbia University's Summer Ecosystem Experiences for Undergraduates Program, an in-depth study-abroad field ecology program. One of the joys of his career has been introducing young children to insects through the Wings, Stings, and Other Strange Things program as well as through the Beetles, Bees, and Bugs courses he has taught at the Natural History Museum at the University of Kansas.

The approaches suggested here are designed to infuse entomology into the classroom, to be a part of the ongoing curriculum or a project. To be successful, though, they do require some preparation. Teachers should read up on insects, particularly those that live in the school neighborhood—ladybugs and other beetles, ants, caterpillars, and bees, for example. Insects are unique in that they are found everywhere; no matter where we live, we can find insects just outside the door, in the shrubs or bushes surrounding the building.

As with any project work, teachers should attempt to involve families in the activities. Invite family members to do some activities at home or to join all the children in a group activity at school. A project-closing celebration that brings families into the classroom conveys a message of importance about the children's work with insects.

## Glossary

**abdomen**—the third and last section of the insect body where the reproductive and digestive organs usually are located

**antennae**—a pair of sensory appendages attached to the head

**camouflage**—the ability to blend in with the surrounding environment

**entomology**—the study of insects

**exoskeleton**—the hard external skeleton of insects (as contrasted with the internal endoskeleton of mammals)

**habitat**—a favorable area in nature where an organism can live and reproduce

**head**—the first section of the insect body, possesses the antennae

**molting**—the shedding of the current exoskeleton and the hardening and expansion of the new, softer exoskeleton that will be larger than the last

**morphology**—the study of insects' structure and function

**predator**—an animal that eats another animal

**specimen**—an organism that has been collected for study

**thorax**—the second section of the insect body, usually includes a pair of wings and three pairs of jointed legs

Introduce scientific terms such as *habitat*; children don't know they are "big" words unless we tell them so, and using the correct terminology makes a child feel more like a scientist.

While introducing children to insects, we should make sure that we model the positive attitudes that we want to develop. Remember, enthusiasm is contagious.

### Live collections

One of the most effective ways of introducing children to the study of insects is by letting them assume the role of entomologist. Rather than gazing at dead insects, children, with our help, can study live specimens. Encourage them to look closely and observe what the insects do and where they live. Don't shy away from scientific terms such as *habitat;* children don't know they are "big" words unless we tell them so, and using the correct terminology makes a child feel more like a scientist. Help the children view insects as living things that eat, sleep, and work. Let them be active investigators who can ask and answer questions about their insect neighbors' lives.

A class collection is appropriate for kindergartners; older children can collect their own insects or form small collecting groups as part of project work. For the initial collection activity, it's a good idea to have all children in the class focus on one species. Pill bugs and ladybugs are ideal first collections because they are slow moving, largely resistant to crushing, large

enough to be well suited for observation, and can be found in abundance among fallen leaves and bushes. Older children may want the challenge of collecting a fast-moving grasshopper. Caterpillars, ants, and flies also make intriguing subjects.

Because these insects are harmless, children can collect them with their bare hands or with a cup. If the insects move too quickly, suggest that the children scoop some dead leaves into a large plastic basin with steep sides and then pick out the insects by hand.

House the insects in clear plastic disposables such as deli or salad containers. Perforating the lids is not necessary; the containers hold more than enough air. But do not add so many insects to a container that they have to crawl over one another.

Once the insects are collected, encourage the children to replicate the habitat in which they were found. Add some leaves for cover. A slightly moistened sponge or paper towel provides a source of moisture. The insects can be kept in the container for a day or two without causing them any harm. Then the children should release them near where they were found.

Teachers should be aware that most children will not want to kill the insects that they collect and that some children will become distressed if the insects are harmed. Handle the *specimens* with care.

Brainstorm ideas with the children about what they want to observe about the ladybugs, pill bugs, or grasshoppers they collect. Help them decide what tools they will need for their observations—magnifying glasses, tweezers, library reference books, and so forth. If, for example, some of the children want to explore what the pill bugs eat, help them plan how to track and report their investigation—documenting it through photographs, graphs, and observation panels.

Small groups might explore different study topics and then report back to the whole class. In this way, children learn a great deal from one another while also becoming experts on one particular aspect of an insect's life. The teacher's role is to facilitate these investigations by steering children to resources in the classroom, at the library, or on the Internet; by planning related field trips; and by helping children learn how to document and summarize what they have learned.

Because such investigations may last a month or more, children can continually collect insects, keeping samples for only a day or two each time. This practice enables the budding scientists to become more confident of their findings because they will observe many different ladybugs or grasshoppers in action. As one study question is answered, the children can pose another one to investigate.

### Build a Bug

An excellent technique to help children understand the rudiments of insect structure and function—the study of which is called *morphology*—is to have them re-create the insect through artistic constructions. From the Reggio Emilia work (Edwards, Gandini, & Forman 1998) we know that children learn best when they can re-create an idea in multiple modalities. Thus, by encouraging children to re-create in a construction what they have observed in real life, we promote learning of the abstract concept.

Have an insect guide on hand so the children can compare the real-life insect with a technical drawing. Any insect guidebook such as the *Peterson First Guide to Insects of North America* (Leahy 1998) will do. Encourage children who are readers to research their insects either at the library or on the Internet. Discuss the pictures with nonreaders.

Children can use any material with which they have had some experience (construction paper, clay, playdough, papier mâché, cardboard, styrofoam, wire, pipe cleaners, and the like). Although children are making a model of a specific insect, artistic license deepens their understanding. The goal is not for the child to create a scientifically accurate rendering of the insect but to construct the insect as he or she sees it, to internalize how it fits together, which features are important, and how the parts relate to the whole. This activity is also a way for children to learn about what is *not* there, as they begin to wonder why the insect does not have eight legs or why it doesn't have bright colors.

Building the insect should not be the final stage of this activity. The children's creations can be useful teaching tools. They can use the models to teach others about the insects. What is its name? Where does it tend to live and on what does it feed? Where are its key morphological components—for example, the *head, thorax, abdomen, antennae* (see "Glossary"), wings, eyes, and mouthparts? These rich vocabulary terms contribute to children's literacy skills as they build scientific skills (Dickinson & Tabors 2001). How many legs does the insect have? How many wings? Can it fly? Do its legs indicate that it can jump like a grasshopper? Does it use its forelegs to grasp other insects like a praying mantis does? What might be the purpose of its coloring or markings: *camouflage* for protection or a warning to *predators* that it doesn't taste good?

To encourage young investigators to be scientifically accurate, teach them correct terms. For example, use the word *antennae* rather than *feelers* (antennae are not used for feeling; they are, instead, primarily used for smell). Build a Bug is the name of this activity because *bug* is a commonly accepted educational term. However, when working with children, try not to use *bug* interchangeably with *insect*. Bugs such as aphids, whiteflies, and mealybugs are a specific group of insects that feed on plants by using piercing, sucking mouthparts.

Finally, make sure that Build a Bug activities are part of children's ongoing project work. When done out of context—for example, as an art exercise—the activity's important potential to educate decreases.

**Children learn best when they can re-create an idea in multiple modalities. Thus, by encouraging children to re-create in a construction what they have observed in real life, we promote learning of the abstract concept.**

## Be a Bee

The Be a Bee activity can help children better understand and recognize insect behavior. It also lets them deepen their appreciation for how insects solve particular problems. Again, although it may be a fun activity in and of itself, it is most meaningful when done in the context of ongoing project work on insects. These activities provide children with another modality (in this case, dramatic play) for learning concepts that they have observed firsthand.

These dramatic-play activities work best when coupled with both direct observation and reading-aloud time. Have on hand insect-oriented books such as Eric Carle's *The Very Hungry Caterpillar, The Very Quiet Cricket, The Very Busy Spider, The Grouchy Ladybug,* or *The Very Lonely Firefly;* Joanna Cole and Bruce Degen's *The Magic School Bus Inside a Beehive;* or Lawrence David and Delphine Durandare's *Beetle Boy.* From an entomologist's viewpoint, *Insects Are My Life,* by Megan McDonald and Paul Brett Johnson, is ideal for introducing children to a wide sample of insect behaviors that can be acted out. First, read about and discuss the behavior. Talk about why it is important. Then let the children loose to act out the behavior.

*The Magic School Bus Inside a Beehive* is an excellent book for introducing bee society and behavior. Young children seem to be universally charmed by the waggle movements of the bee dance. The waggle dance, shown on page 37, is used by bees to communicate distance, direction, and the quality of the food they have found to the other bees back at the hive (Von Frisch 1993). The dance is performed with a series of repeated figure eights. The center of the figure eight is where the bee sends its message. The orientation of the center line communicates the direction of the food source relative to the sun. In the illustration, the bee dances vertically up the comb, showing that the food source is straight into the sun. (If the food source were, instead, located 90 degrees to the right of the sun, the waggle would be oriented toward the three o'clock position.)

The length of the waggle represents distance. The intensity indicates the desirability of the food source—the wilder the waggle, the better the meal. A group of children acting out this dance, with everyone leaning forward and shaking his or her derriere like a bee, can bring even a serious visiting scientist to his knees in laughter. To deepen understanding, ask the children to do the dance right before snacktime to signify where the snack table is and just how much they like the treat.

Equally thrilling to most children is acting out *molting*—the insect developmental process whereby the old *exoskeleton* (sometimes called *skin*) is shed to allow for growth. The exoskeleton of insects is very hard and protects the delicate inner body parts. It also restricts the maximum body size of the insect and must be shed before the insect can proceed to the next developmental step. Many children have some understanding of caterpillars and know that they change into butterflies or moths when they reach adulthood. Indeed, they may have observed this phenomenon firsthand by collecting caterpillars for project work. *The Very Hungry Caterpillar* can build children's knowledge and inspire their imaginations. After reading the book, explain that an insect needs to shed its skin when it gets too big for it because its skin (exoskeleton) does not expand like ours does. We can make our skin larger, as we can see when we pinch it, but the caterpillar's skin is hard and cannot be pulled. Before shedding, the caterpillar produces a new, larger skin underneath. The caterpillar must then crack open its old exoskeleton and push its body out of the crack. Once the creature leaves its old skin behind, it needs to gulp lots of air to expand the new exoskeleton to as large a size as possible. It must retain the air until the new exoskeleton dries and hardens because, from this point on, the exoskeleton will not get any larger.

Most insects go through several molts during their lifespans; some undergo more than 20 of these changes! Encourage children to act out the caterpillar's molting process. As a follow-up, children could draw or paint the processes they have acted out or set their dramatizations to music. Biology readily merges with the arts in this activity.

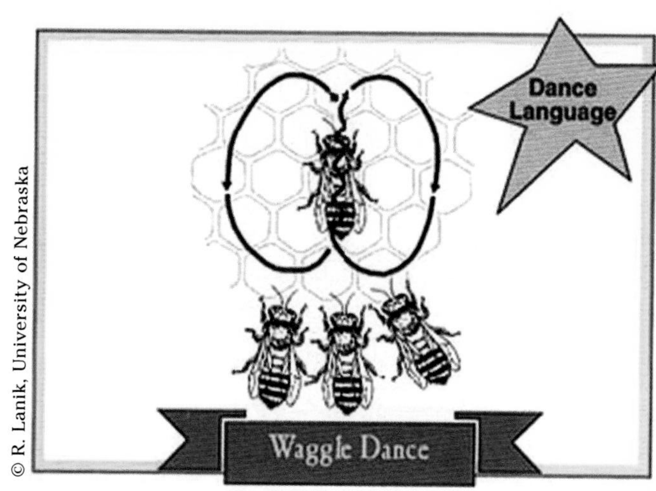




The waggle dance is used by bees to communicate distance, direction, and the quality of the food they have found to the other bees back at the hive.

An insect needs to shed its skin when it gets too big for it be- cause its skin (exoskeleton) does not expand like ours does.

## Conclusion

An active, hands-on approach to science makes entomology both exciting and relevant to young children. Through a positive introduction to insects, children can develop the observation and analysis skills that will serve them well throughout their academic schooling and later life. Providing children with a positive introduction to entomology enables them as adults to understand the importance of biodiversity and to be both comfortable with and respectful of insects.

## References

Dickinson, D.K., & P.O. Tabors, eds. 2001. *Beginning literacy with language.* Baltimore: Paul H. Brookes.

Edwards, C.P., L. Gandini, & G.E. Forman, eds. 1998. *The hundred languages of children: The Reggio Emilia approach—Advanced reflections.* 2d ed. Norwood, NJ: Ablex.

Leahy, C. 1998. *Peterson first guide to insects of North America.* Boston: Houghton Mifflin.

Lind, K.K. 1999. Science in early childhood: Developing and acquiring fundamental concepts and skills. In *Dialogue on early childhood science, mathematics, and technology education.* Washington, DC: American Association for the Advancement of Science.

Von Frisch, K. 1993. *The dance language and orientation of bees.* Cambridge, MA: Harvard University Press.

Wilson, E.O. 1992. *The diversity of life.* New York: W.W. Norton.

# Raising Butterflies from Your Own Garden

Patricia Howley-Pfeifer

Children love nature and, in particular, butterflies. Older preschoolers and primary age children learn in important ways from the hands-on experiences of raising butterflies. Even if your school is in the city, planters (or a garden) filled with caterpillar-host and nectar flowers specific to your region will definitely attract these beautiful creatures. A butterfly landing in a garden or on plants cared for by the children generates much excitement, promoting feelings of pride, protection, and guardianship.

Garden butterflies do not scare off easily. Often, when butterflies such as monarchs, black swallowtails, tiger swallowtails, and some skippers are sipping nectar, you can easily observe from within a space of two feet. I strongly endorse the North American Butterfly Association's belief in observing butterflies through naked-eye observation or with binoculars, not capturing them. Children and adults can have excellent observation experiences simply by standing next to plants such as a butterfly bush or a New England aster.

> Children can have excellent observation experiences simply by standing next to plants such as a butterfly bush or a New England aster.

Through these experiences, children discover that nature can be studied and enjoyed without impact or interference from observers. Because monarch, black swallowtail, and mourning cloak butterflies are found throughout the country, I have chosen to focus this article on these three varieties.

---

**Patricia Howley-Pfeifer**, M.A., kindergarten teacher at Winfield School in Winfield, New Jersey, has been an early childhood teacher for 24 years. Patricia is very involved in environmental education and has won the Environmental Education Award for Excellence from the New Jersey Audubon Society.

Photos courtesy of the author.

## How to raise a butterfly

I rely on the caterpillars I find in my home and school garden rather than use commercially grown caterpillars. (See the North American Butterfly Association Website at www.naba.org for more information about commercially grown caterpillars.) Finding eggs and caterpillars and then raising them in the classroom can be a fun and rewarding experience for children, families, and teacher, and finding these living things exactly where nature placed them actively engages the children in the discovery and investigation process of science.

The following words of advice will be useful to anyone planning to raise and observe butterflies.

### Before you begin

**1.** Practice raising caterpillars at home before involving children. To attract monarchs, plant milkweed (any *Asclepias* variety) in your garden or in a planter. Black swallowtails prefer curly parsley, and mourning cloaks favor willow trees.

**2.** Gather clear containers to house eggs, larvae, pupae, or butterflies. Large pickle jars, large pretzel containers, or fish tanks are safe and can provide space. Make sure there is an air supply and keep the containers out of direct sunlight. Secure some type of netting or screen over the mouth of the container so caterpillars and butterflies do not get out.

**3.** Become familiar with the "Stages of Growth" section of this article. Then you will know what to expect during the different stages.

**4.** Plan to provide daily care. Raising caterpillars is not difficult, but it is time-consuming. Until caterpillars reach the chrysalis stage, they will not do well if left over the weekend in the classroom.

**During the growth process**

**5.** Monitor the eggs and caterpillars. After an egg hatches, no new leaves will be needed for a day or so unless the plant the egg was on has wilted. When adding new sprigs with leaves, place them close to the old sprigs in a vase of water within the container so the small caterpillars can move easily to the new food. When the caterpillars are very small, place one to two sprigs per caterpillar in the vase. Remove the old sprigs only after the caterpillars have moved off them. As the caterpillars mature, three or more sprigs per caterpillar per day will be needed, depending on how much they eat. The larger the caterpillar, the more it will eat. Change the water in the vase when new food is added. If you buy parsley, be sure that it is truly organic; chemicals on store-bought parsley can produce immediate deadly results.

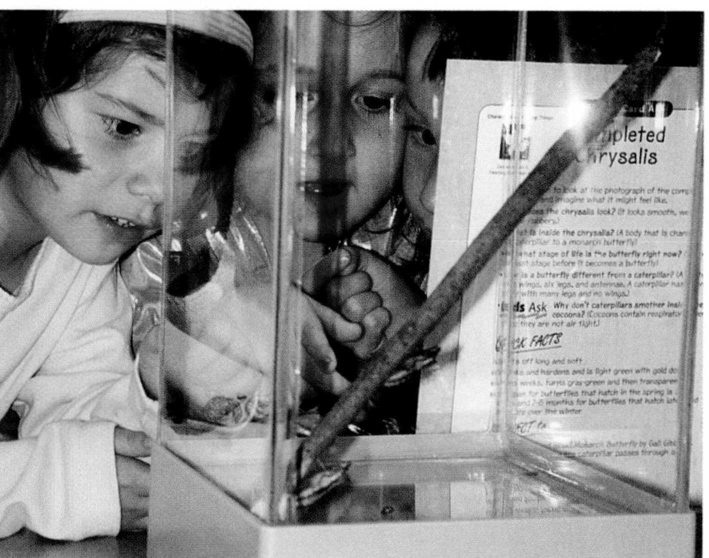

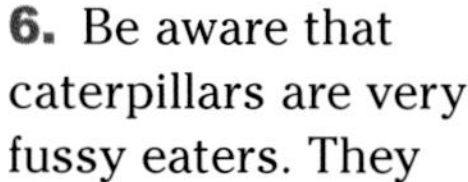

**6.** Be aware that caterpillars are very fussy eaters. They will eat only the leaves of their specific host plant. For this reason, the female butterfly will lay the eggs only on the host plant—to ensure that her larvae will survive. Some butterflies, such as monarchs, obtain lifesaving toxins from the leaves they eat, an insect–plant adaptation that your class can investigate.

**7.** Keep the caterpillar's dwelling clean. Remove wilted leaves and caterpillar droppings (frass) to prevent diseases and mold growth. Keep the container free of parasites. Spiders and other insects must be removed (by hand, not with chemicals), or they may eat the caterpillars.

Finding these living things exactly where nature placed them actively engages the children in the discovery and investigation process of science.

**8.** Make sure that caged caterpillars have a place to become a chrysalis. Some of the caterpillars we raised attached themselves to the side of the fish tank they were in, and a few hung from the screen covering the tank. The black swallowtails and mourning cloaks attached themselves to the sturdy sticks we placed in the tank. Release butterflies within a day of emergence from the chrysalis, unless it is raining.

**9.** Look for shed skin on the bottom of the container after the caterpillar molts into a chrysalis. Put the skin and empty chrysalises in small magnifying containers to allow further observations. Many nature centers can supply these containers.

**Make it a learning experience**

**10.** Help children understand that handling the caterpillars or butterflies may accidentally harm the animals. Children benefit from realizing that they can learn about and love animals without handling them.

**11.** Keep detailed records of your experiences in raising butterflies. These records help you and the children to remember mistakes, successes, growth rates, and other information.

## Stages of growth

The sections that follow describe the stages of growth that occur in the development of a butterfly. Suggestions about what to have children look for are also included.

### Egg

The female butterfly searches for the host plant by tasting the plant with her feet. Some butterflies, such as monarchs and black swallowtails, lay a single egg at a time. The female bends her abdomen into a **C** shape to lay the eggs directly on the plant. (Monarchs lay eggs on the underside of leaves. Black swallowtails lay directly on the parsley.) The eggs take about three to four days to hatch.

### Caterpillar (larva)

A tiny caterpillar, or larva, hatches from the egg and enjoys its first meal—its egg case! The size of the caterpillar determines how much it eats and how active it is. Have children note eating habits and observe whether some caterpillars are more active

than others. Does the rate of consumption change as the caterpillar grows? Also, children can observe eating behaviors, such as the direction and pattern of eating, how many leaves are eaten in a certain period of time (e.g., over several days), and whether all the caterpillars (or even an individual caterpillar) follow the same eating routine.

Caterpillars grow and change (molt) a number of times during their lifetime. Many caterpillars molt four to five times. When the caterpillar grows too big, the old skin is shed. Children can notice, observe, draw, and describe these stages in their journals. Each child can study and compare one caterpillar with another.

The caterpillar's colors (e.g., black swallowtail caterpillars change colors as they grow) and body structure (e.g., some look like they have two heads) are essential for survival. Lead the children in a discussion about the reasons for camouflage. Pictures of other animals using camouflage can spark investigations. I developed a short puppet show using praying mantis, ladybug, and butterfly puppets to demonstrate toxic coloring and camouflage.

When a caterpillar is ready to begin the final stage of development, it expels (throws up) excess fluid and starts to look for a safe spot to pupate (become a chrysalis). The monarch caterpillar looks for a safe location such as the top of the container and attaches a white silk mat that is produced from a gland near the mouth to the chosen location. Then it attaches small hooks located near the end of its abdomen to the silk pad and hangs upside down. Slowly, its body forms a **J** shape, an indication that the larva is getting ready to pupate. The time from egg to chrysalis varies from caterpillar to caterpillar, and it varies with the temperature. Have the children compare and graph the amount of time it takes the egg to hatch, how long the caterpillar phase lasts, and the number of days from egg to chrysalis.

### Chrysalis (pupa)

A moth larva produces a cocoon by spinning itself into a silk enclosure. Butterflies do not make cocoons. Instead, they pupate from the caterpillar stage into the chrysalis stage. You and the children can explore the chrysalis's use of camouflage.

About a day before the butterfly emerges, the chrysalis begins to lose its color. By the time the butterfly emerges, the chrysalis will become transparent, and you will be able to see the butterfly all squished up in the chrysalis. (A mourning cloak chrysalis does not become transparent, but the chrysalis shakes just before the butterfly emerges.) Emergence takes about one minute. The chrysalis cracks, and the butterfly emerges—all curled and crunched up. The butterfly immediately grabs onto the empty chrysalis with its feet and stays there for several hours. In time, the butterfly will exercise its wings before attempting to fly. During the chrysalis stage, the children can graph the number of days each animal was in a chrysalis, when the chrysalis started to become transparent, and the time of day each butterfly emerged. Excitement will mount as the times for emergence draw near!

Adult butterflies do not continue to grow; the adult stage is the final phase. The time from egg to chrysalis is about 17 to 24 days. Caterpillars develop faster in warmer summer temperatures and slower during cool fall weather. With the children, research how temperature affects growth and development.

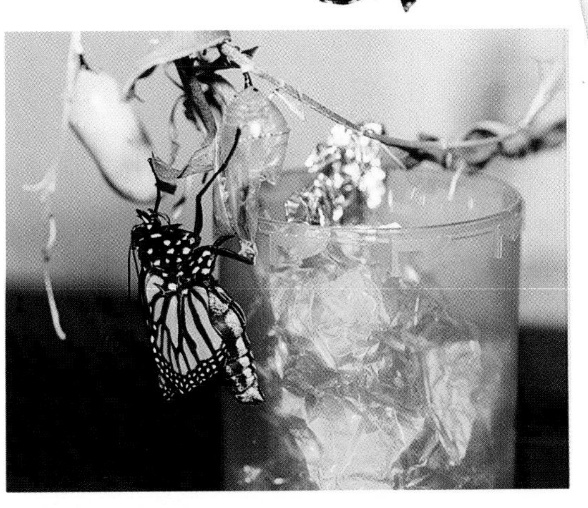

We saw the butterflies fly away. there was four of them. Some were on the top of the cage. We let them go outside. the butterflies are going to Meyico.
Amber

Tyler
First the chrysalis was green and then it was black It had golden lines on it. the butterflies were orange and black.

## Classroom activities

Children can document the growth of the caterpillar in a dated journal. Teachers and children can take photographs of the caterpillar's growth and, at times, of the children with the caterpillars, chrysalises, and butterflies. If children are fortunate enough to observe molting, pupating, or butterfly emergence, have them write about and draw pictures of the transformation immediately afterward. Teachers can record children's dictated words.

Encourage the children to ask questions. Don't be afraid to say you don't know an answer. You can teach the children research skills as, together, you look for answers on the Internet or in books. A teacher's sense of awe and excitement about butterflies and about the whole learning process can encourage and facilitate children's love of nature. Also, in our classroom, I have a variety of adult field guides and nature books (borrowed from libraries or bought at garage sales) that the children use during free book time.

During our butterfly research, we have many discussions about the colors, stripes, and patterns on the caterpillars and butterflies. Our class butterfly book includes the children's observations; drawings; dictations; and photographs of the children with the butterflies, caterpillars, and chrysalises. In particular, children learn how to observe and communicate logical thinking related to their observations by looking (with and without a hand lens) at the egg, caterpillar, chrysalis, and butterfly and by drawing what they see and then writing or dictating their observations. The teacher supports the discovery

> A teacher's sense of awe and excitement about butterflies and about the whole learning process can encourage and facilitate children's love of nature.

process by helping children to communicate what they are seeing and thinking. The following are examples of dictations that children made about our butterflies:

> As each monarch flew off, we waved and said, "Good-bye! Good luck! Have a nice time in Mexico!"

**Ciara**—I made the (black swallowtail) caterpillars hanging onto the stick. I saw the butterfly laying the eggs. The butterfly was blue and yellow and black. It was drinking nectar from the purple flowers. The caterpillars have stripes on them. They hang from the string on the stick.

**Casey**—The caterpillars are in the chrysalis. We drew pictures of them. I drew the stick in the tank. I drew the skin, and the chrysalis has spikes and string. The butterfly is going to come out in the spring. The chrysalis is camouflaged to the wood so animals won't try to eat it, because they will think it's part of the wood. The chrysalis is attached to the wood, and they can't see the string because it's so tiny and they think it is part of the wood.

**Tyler**—The chrysalises are on a twig. There are two strings attached to it. There are spikes on it. The little black stuff on the bottom is the skin of the caterpillars when they went into the chrysalis. The caterpillars are going to turn into butterflies.

**Tommy**—I like all of them (mourning cloaks). They have spikes. They eat and get big. If I were them, I would ask them to call me Stuffer Face because they stuff themselves with leaves. My favorite part is when they eat. They are amazing! I like how they bunch up on the branches to scare away the birds, so they look like one big caterpillar so they don't get eaten. I bet they can really trick animals because they fake a head in the back of them and when the enemy comes it will pull out its tail instead of the head. I like the small ones because they are like me.

In addition, through creative dramatics, we act out the butterfly's life cycle and other aspects of its life; we form our bodies into the appropriate stage, and during the acting, we talk about what we are experiencing. Younger children love to use a high voice as they act out a butterfly feeling squished and wanting to come out of the chrysalis.

Finally, involving families in the children's nature lessons strengthens learning for their child. The adult and child share a common knowledge, and both can expand on it or reinforce the concepts learned. On one occasion, family members came to our classroom and did an art project about butterflies with their child. Later, they came to our Monarch Butterfly Release Party. At the end of the school day, I took the monarch container outside and allowed parents to have some observation time. When I came out with the children, I recapped our classroom experience and shared the fact that the monarchs would be migrating to Mexico when they were released. As each monarch flew off, we waved and said, "Good-bye! Good luck! Have a nice time in Mexico!"

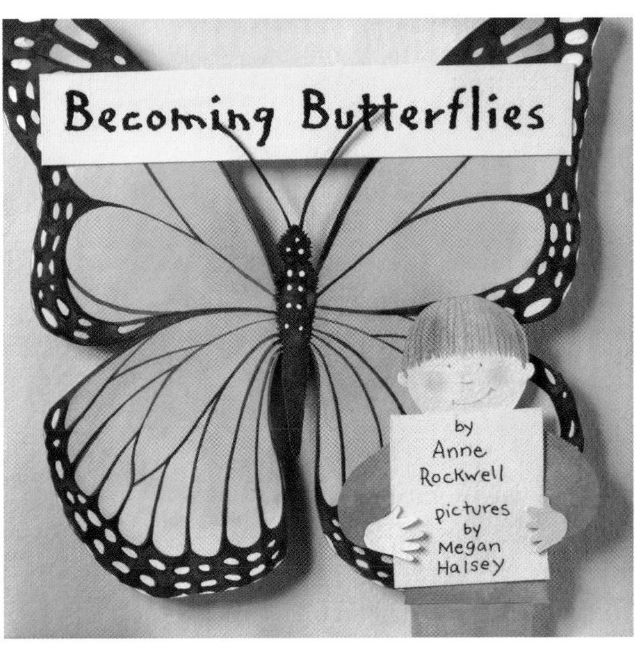

# Print and Online Resources That Spotlight Young Children and Science

### compiled by Alice Galper and Carol Seefeldt


Agler, L. 1990. *Liquid explorations.* Berkeley, CA: Lawrence Hall of Science.

American Asssociation for the Advancement of Science. 1989. *Science for all Americans.* New York: Oxford University Press.

American Association for the Advancement of Science. 1993. *Benchmarks for science literacy.* New York: Oxford University Press.

American Association for the Advancement of Science. 1998. *Dialogue on early childhood science, mathematics, and technology education.* Washington, DC: Author.

Arnosky, J. 2002. *Field trips.* New York: HarperCollins.

Barclay, K., C. Benelli, & S. Schoon. 1999. Making the connection! Science and literature. *Childhood Education* 75: 215–24.

Bredekamp, S., & T. Rosegrant, eds. 1995. *Reaching potentials: Transforming early childhood curriculum and assessment.* Vol. 2. Washington, DC: NAEYC.

Burton, R., & S.W. Kress. 1999. *The Audubon backyard birdwatcher: Birdfeeders and birdgardens.* San Diego: Thunder Bay.

Chalufour, I., & K. Worth, for Education Development Center, Inc. In press. *The Young Scientist Series.* St. Paul, MN: Redleaf Press. ["Discovering Nature with Young Children," "Exploring Water with Young Children," and "Building Structures with Young Children"]

Desrochers, J. 2001. Exploring our world: Outdoor classes for parents and children. *Young Children* 56 (5): 9–12.

DeVries, R. 2002. *Developing constructivist early childhood curriculum.* New York: Teachers College Press.

Donoghue, M.R. 2001. *Using literature activities to teach content areas to emergent readers.* Needham Heights, MA: Allyn & Bacon.

Fleer, M., & A. Cahill. 2001. *I want to know . . . ? Learning about science.* Watson, Australia: Australian Early Childhood Association.

Gallenstein, N.L. 2003. *Creative construction of mathematics and science concepts in early childhood.* Olney, MD: Association for Childhood Education International.

Galvin, E.S. 1994. The joy of seasons: With the children, discover the joys of nature. *Young Children* 49 (4): 4–9.

Gartrell, J.E., J. Crowder, & J.C. Callister. 1992. *Earth: The water planet.* Arlington, VA: National Science Teachers Association.

Glassberg, J. 1997. Release that net: You need two hands for binoculars! *American Butterflies* (Spring): 2–3.

Glassberg, J., P. Opler, R.M. Pyle, R. Robbins, & J. Tuttle. 1998. There's no need to release butterflies—They're already free. *American Butterflies* (Spring): 2–3.

Gochfeld, M., & J. Burger. 1997. *Butterflies of New Jersey.* New Brunswick, NJ: Rutgers University Press.

Harlan, J.D., & M.S. Rivkin. 2000. *Science experiences for the early childhood years: An integrated approach.* 7th ed. Upper Saddle River, NJ: Merrill/Prentice Hall.

Jackson, D. 2002. *The bug scientists.* Boston: Houghton Mifflin.

Kneidel, S. 1993. *Creepy crawlies and the scientific method: More than 100 hands-on science experiments for children.* Golden, CO: Fulcrum.

Kramer, D. 1989. *Animals in the classroom: Selection, care, and observations.* Menlo Park, CA: Addison-Wesley.

Lind, K.K. 2000. *Exploring science in early childhood education: A developmental approach.* 3d ed. Albany, NY: Delmar.

Lowery, L., ed. 1997. *NSTA pathways to the science standard: Elementary school edition.* Arlington, VA: National Science Teachers Association.

Martin, D.J. 1996. *Elementary science methods: A contructivist approach.* Arlington, VA: National Science Teachers Assocation.



**Alice Galper**, professor of curriculum and instruction at the University of Maryland, has authored many books and articles.

**Carol Seefeldt**, professor emerita at the Institute for Child Study, University of Maryland, and visiting scholar at Johns Hopkins University, has taught at every level from preschool though grade three and written numerous books, articles, and pamphlets.



McKissack, P.C., & F. McKissack. 1994. *African American scientists.* Brookfield, CT: Millbrook.

National Geographic. 2000. *National Geographic animal encyclopedia.* Washington, DC: Author.

National Geographic. 2002. *Honeybees.* Washington, DC: Author.

National Research Council. 1996. *National science education standards: Observe, interact, change, learn.* Washington, DC: National Academy Press.

Paulu, N. 1992. *Helping your child learn science.* Washington, DC: U.S. Department of Education, Office of Educational Research and Improvement.

Pringle, L. 1977. *Extraordinary life of the monarch butterfly.* New York: Orchard Books.

Rivkin, M.S. 1995. *The great outdoors: Restoring children's right to play outside.* Washington, DC: NAEYC.

Seefeldt, C., & A. Galper. 2002. *Active experiences for active children: Science.* Upper Saddle River, NJ: Merrill/ Prentice Hall.

Sussman, A. 2000. *Dr. Art's guide to planet Earth: For earthlings ages 12 to 120.* White River Junction, VT: Chelsea Green.

Sutton, P. 1998. How to create a butterfly and hummingbird garden. www. njaudubon.org/naturenotes/garden.html.

Torp, L., & S. Sage. 1998. *Problems as possibilities: Problem-based learning.* Washington, DC: Association for Supervision and Curriculum Development.

Wenner, G. 1993. Relationship between science knowledge levels and beliefs toward science instruction held by preservice elementary teachers. *Journal of Science Education and Technology* 2: 461–68.

Wick, W. 1997. *A drop of water: A book of science and wonder.* New York: Scholastic Trade.

Wilson, R.A. 1993. *Fostering a sense of wonder during the early childhood years.* Columbus, OH: Greyden.

Wilson, R.A. 1995. Nature and young children: A natural connection. *Young Children* 50 (6): 4–7.

Worth, K., & S. Grollman. 2003. *Worms, shadows, and whirlpools: Science in the early childhood classroom.* Portsmouth, NH: Heinemann; and Washington, DC: NAEYC.


## Online resources

**www.ala.org/parentspage/greatsites/science.html/**. Created by the Howard Hughes Medical Institute, this site offers both online and offline science activities for children of all ages.

**www.allaboutbutterflies.com**. All About Butterflies

**www.drscience.com**. Make science inquiries and receive a daily science question from Dr. Science.

**www.edna.edu.au/schools/earlychildhood/index.html/**. This Australian site, What's New at EdNA Online for Early Childhood, provides educators with links and e-mail about science, including *ABC Science: Walking with Beasts* and *CSIRO Science by Email.*

**www.exploratorium.edu/**. San Francisco's Exploratorium, the museum of science, art, and human perception, offers virtual tours of exhibits and resources for science educators.

**www.garden.org**. National Gardening Association

**www.geocities.com/ciarasbutterflies**. Ciara's Butterflies

**www.monarchlab.umn.edu**. Monarch Lab Exploring Monarch Butterfly Biology. St. Paul: University of Minnesota.

**www.mos.org/**. Boston's Museum of Science provides an interactive look at its exhibits.

**www.naba.org**. North American Butterfly Association

**www.nsta.org/**. The National Science Teachers Association offers a wealth of online information on teaching ideas.

**www.nwf.org**. National Wildlife Federation

**www.pbs.org/teachersource/sci_tech.htm/**. This PBS site provides preschool and primary educators with activities and ideas for teaching science.

**www.project2061.org**. Project 2061, a long-term initiative of the American Association for the Advancement of Science, works to reform K–12 science, mathematics, and technology education.

**http://sln.fi.edu/educators.html/**. Philadelphia's Franklin Institute Online provides educators with science lesson plans, activities, and virtual tours of activities.

**http://yucky.kids.discovery.com/flash/index.html/**. The Yuckiest Site on the Internet, operated by the Discovery Channel, teaches children science through interactive games. Though it's intended for children, who could resist "A Day in the Life of Ralph Roach"?

Reflecting, Discussing, Exploring

# Questions and Follow-Up Activities

The articles in *Spotlight on Young Children and Science* represent just a small sample of the many valuable resources for early childhood educators interested in science for young children. For students in early childhood professional preparation programs, for early childhood teachers taking part in training and other forms of professional development, and for individuals seeking to broaden understanding of this important curriculum area, we hope these articles and the accompanying professional development resources (opposite) will open doors to further work in this exciting area.

To help you reflect on and apply insights from these articles, we have developed a series of questions and suggested follow-up activities. The series begins with an invitation to think about your own early experiences with science. Specific questions and suggested activities related to each article follow. Finally, we help you pull things together with general questions about curriculum, teaching practices, resources, and next steps.

## A. Recalling your own early experiences

1. What memories do you have of early experiences learning about science in school? What beliefs and attitudes grew out of these early experiences?

2. Outside of the classroom, what do you remember about exploring the world of everyday scientific phenomena—for example, puddles, grass growing up through a pavement, dirt, reflections, shadows? Was there something special that intrigued you? Why did you find this interesting?

## B. Expanding on each article

**"Science in the Preschool Classroom: Capitalizing on Children's Fascination with the Everyday World to Foster Language and Literacy Development"**

Kathleen Conezio and Lucia French describe the benefits of a coherent, hands-on science curriculum for young children—including its value for language and literacy learning.

1. The article begins with the statement "A young child starting preschool brings a sense of wonder and curiosity about the world." See if you can identify this sense of wonder and curiosity in action, and discuss how you might extend or build on those feelings.

2. The authors assert that three key components need to be included in early childhood science—content, process, attitude. Using an everyday experience in your classroom (e.g., washing tables after snack), see if you can identify places where you could build greater focus on scientific content, process, or attitude.

3. The authors' discussion of "coherence" is very important, and not just for science. Develop a checklist of questions to ask when determining whether a curriculum or environment is "coherent." Take an activity or experience that's currently in your program and suggest two or three ways to deepen it by making the experience more coherent.

4. If your program has specific language and literacy goals or standards, consider how some of these could be supported through science activities and explorations. Also consider how literacy experiences (e.g., reading nonfiction and even fiction books relating to science) may, in turn, promote science learning. Might these literacy connections also apply to mathematics?

**"Entries from a Staff Developer's Journal . . . Helping Teachers Develop as Facilitators of Three- to Five-Year-Olds' Science Inquiry"**

Robin Friedrichs Moriarty describes how she and colleagues helped Head Start teachers become more confident in supporting children's scientific inquiry through their own reflection and exploration.

5. Teachers need a foundation of content knowledge in science in order to help children learn. The author describes how she brought teachers together to do their own water investigations as a way to build this knowledge. How effective do you think this approach would be? What else might be needed?

6. Try getting involved in simple water play—alone or with fellow teachers. If you do not have the equipment described in the article, improvise your own. What do you learn about water, water flow, and so on? How might this help you plan good science experiences for children?

7. If you work with teachers or student teachers, try keeping your own reflective journal as this staff developer did, even for a few weeks. How does this help you focus on staff or student growth and on your role in supporting that growth?

**"Using Photographs to Support Children's Science Inquiry"**

Head Start teacher Cynthia Hoisington tells of her use of digital photographs to challenge children to extend their science investigations and learning.

8. Explore the fascination and power of photography to explore what children are learning in science and other areas. Whatever kind of camera you have, begin or expand your photography. Alone or with a group, create a display with brief descriptions of what children are learning. Use these displays to remind children of their learning and to communicate with families and staff.

9. Using photos to document young children's activities, and to support teachers' reflection, is a valuable technique—including and beyond the science arena. Try creating posters or documentation panels like those described in this article—on this or other topics. (The book *Windows on Learning,* mentioned in the article, would be a helpful resource.)

10. Select another activity that children are engaged in, and plan how to use photos to support their exploration and learning.

11. This teacher used still photos. What difference in the learning possibilities might you expect if you videotaped children's science explorations/investigations rather than taking photographs? For what kinds of science experiences or what kinds of learning objectives might each be preferable?

**"Documenting Early Science Learning"**

How do you know what young children know? Authors Jacqueline Jones and Rosalea Courtney suggest classroom-based approaches to gathering evidence of children's growing understanding of science.

12. Review the three principles described in this article. Why are they so important? What do you need to do to enhance your capability to address these principles?

13. Explore some of the broader issues in early childhood assessment using the references in this article, in the "**Resources**" section (p. 44), in NAEYC position statements, and elsewhere.

14. Think about the fascinating topic of children's "wrong answers." In many of the classroom conversations recorded in this article, children in kindergarten and first grade—and adults, for that matter—have incorrect ideas about basic scientific phenomena. How important is it to correct them? What can teachers learn if they do not correct these ideas immediately, and how else might children arrive at better understanding?

15. Use the "Steps in Guiding a Discussion of Children's Language Records" process described in this article to gain insight into children's thinking and learning about science and other topics.

**"Be a Bee and Other Approaches to Introducing Young Children to Entomology"**

Scientist James A. Danoff-Burg offers scientifically sound yet engaging and creative approaches to introducing young children to a specific science topic, the study of insects.

16. What tips in this article (e.g., representation, depth of study, active teacher involvement) reflect and support those in any of the other articles?

17. The focus of this article is on the kindergarten and primary grades. What adaptations might be needed if you were to investigate insects with younger children? How and why would these changes support younger children's exploration and learning?

18. The author suggests that insect study can support other areas of learning, including language

development, while at the same time building scientific concepts. You might explore these possibilities in more depth, referring to standards for early language, literacy, and mathematical development.

**"Raising Butterflies from Your Own Garden"**

Kindergarten teacher and environmental educator Patricia Howley-Pfeifer shares the excitement, learning opportunities, and practical how-to's of raising butterflies as a class project.

19. What basic principles of good curriculum planning are represented here? How could you use these to investigate a different topic?

20. Although this project has a science focus, it supports learning in many other areas as well. What are some of these areas?

21. What goals were accomplished through approaching the study of butterflies as an extended project rather than simply as a theme or one-time "lesson"? For more information about projects, you might look at *Young Investigators,* by J.H. Helm and L. Katz (NAEYC and Teachers College Press, 2001) or the Project Approach Web site at www.project-approach.com.

22. The author and teacher in this article shared her own passion for butterflies with the children in her class. What special interest of yours could you share with children? How might it support their learning about science or in other areas?

## C. Making connections

### Consider the big picture

1. What, in your view, are the three most important themes or key ideas that recur across this group of articles? If possible, compare your nominations with those identified by other readers.

2. Again thinking of the entire group of articles, what are three key teacher behaviors that support young children's science learning? And what are three aspects of the classroom environment that do the same?

3. This is a small number of articles, and so some important ideas surely have been left out. What, in your mind, is missing in these discussions of early childhood science? For example, family engagement? children with disabilities? links between mathematics and science? others? Where might you learn more about these missing pieces?

### Examine curriculum goals and expected outcomes

4. Are there some key concepts and understandings (big ideas) about science that children should learn in preschool? kindergarten and first grade? How should this be determined?

5. The introduction to this set of articles (pp. 2-3) cites a number of science goals and benchmarks. Examine these and discuss in a group, perhaps having group members choose certain benchmarks to present to the rest of the group. How appropriate are these benchmarks as guides for your work? How might they influence your curriculum and teaching practices?

6. Reports from Project 2061 of the American Association for the Advancement of Science and from the National Research Council are at the core of the suggestions in these articles. With colleagues or fellow students, share responsibility for reading these important reports and summarizing their findings. How well do you think your program, or the early childhood field, is doing with respect to the key recommendations made in those reports?

7. What other science goals and standards would be useful to examine? Many states have developed content or outcome standards for children below kindergarten; some of these include science (visit, e.g., www.educationworld.com/standards/). Also consider reviewing the Head Start Child Outcomes Framework's indicators in the Science domain (online at www.hsnrc.org/hsnrc/CDI/pdfs/Outcomesbroch.pdf). Working with colleagues or fellow students, compare these with national goals and benchmarks—what are similarities and differences in their expectations?

### Use reflection to enhance teaching practices

8. Several of the articles have good examples of teacher questions that challenge children's thinking. What characteristics do these questions have in common? How can you increase or improve your use of questions to encourage children to think in deeper ways?

9. Many of these articles describe classrooms in which teachers and children are engaged in deep, sustained scientific learning. For such engagement to occur, what supports and program characteristics need to be in place?

10. As you read and discuss all the articles, what do you find that affirms your current practices? What questions do these articles raise about your current practices? What new approaches might you try?

11. What role does children's direct exploration and "messing about" with materials play in their science learning? When is it—and when is it *not*—a waste of time?

12. How might you use photos of individual children's work to encourage reflection and extension—in science and in other areas? What makes photography such a powerful technique? What other forms of documentation have strong potential, and for what purposes?

13. Select one or more of the photos from the articles. Develop a set of questions to help you and your colleagues probe the issues behind these photos in more depth. In one photo, for example, a child seems deeply absorbed in scooping dirt into a cup. What questions might be in her mind? What might her teacher do to support further investigation?

14. Most early childhood settings include children for whom adaptations or accommodations are needed in order to respond to and build on individual interests, talents, abilities, languages, cultures, and more. Identify one child you teach and consider how you might alter some aspects of the environment, activities, interactions, family communication, and so on to help that child benefit from the approaches to early childhood science learning described in these articles.

### Focus on families and communities

15. What strategies are described in these articles that can help communicate to families what their children are doing and learning in the area of science? What other strategies or variations can you think of?

16. How might you engage families in encouraging children's exploration of and learning about science?

17. Consider developing a plan for a "Family Science Night" where families could learn about and actively participate in their children's science curriculum.

18. Most families are legitimately concerned with developing their children's literacy skills. How might you communicate with families about the important connections between science and literacy, or between mathematics and literacy, in a jargon-free and interesting way?

19. All families love to see photographs of their own children. How could you use photos like those in these articles to help family members experience the wonder of their children's science learning? Try adding captions or "thought balloons" to photos and creating displays or documentation panels. (Here again, *Windows on Learning* would be helpful.)

20. Every community is different, yet each offers many opportunities for everyday scientific investigations. Identify one such opportunity in the community where you live and work. How might you use that to extend children's science learning?

### Identify resources and plan next steps

21. The "**Resources**" section (p. 44) contains a rich menu of science-related books, articles, and Web sites. In addition, most articles in this cluster have a list of references specific to that topic. Select one or more of these resources and write an annotated description of it to guide others—perhaps putting this information in handouts or on a Web page. What is the early childhood content of the material? For which professionals is it especially valuable? For which children?

22. Besides those in this book, what other resources have you found to support your work on early childhood science? Again, you might create an annotated list to share.

23. Develop rating scales or criteria (e.g., use of current, research-based information; attention to issues of culture, language, individual differences; attention to content, processes, and attitudes) that could be used to assess the value of such resources.

24. What materials and resources do you think are necessary for a high-quality early childhood science program, and how could these be acquired? Make a list, with estimated costs and sources, including possibilities for free or donated materials.

25. What do you feel you need to know more about in order to have a more effective science program? What will you want to change (in your classroom, your schedule, some aspects of teaching practices)?

26. Develop specific plans to engage the children in your class more fully in science inquiry. Create an action plan to guide this work. Implement your plans, and record what happens through observation notes, journal entries, video, or photos.